A FALCON ʄIELD GUIDE

the Great Lakes Berry Book

A complete guide to finding, harvesting, and preparing

wild berries and fruits in the Great Lakes.

Includes 115 recipes and 48 color photos.

B o b K r u m m

FALCON

Helena, Montana

Front cover photo: David Papazian

All photos in the color section are those of the author, unless otherwise noted.

Library of Congress Catalog Card Number: 96-85106

Printed in Canada.

♻ Text pages printed on recycled paper.

CONTENTS

ACKNOWLEDGMENTS

A host of people helped me in one way or another to write this book. First and foremost I want to thank my parents, Donald and Emily Krumm, for all they taught me—from identifying plants, to gardening, to cooking. Their instruction and love have sustained me well over the years.

Doctors E. A. Stowell and Dan Skean of Albion College in Albion, Michigan, supplied many of the photos I didn't have. Dr. Stowell nurtured my joy of learning and taught me a scientific, systematic way to identify wild plants.

Thanks to my helpmate Dot Heggie for encouraging me to keep on when I would rather have quit. She prodded me to start writing the Berry Book series. Dot also served as proofreader and test cook.

Ted and Toni Terrel supplied much of the information on the Persimmon chapter. Their close friendship for over thirty years has been a source of joy and warmth.

Sharon Henry, Rose Henckel, the Lewistown Chamber of Commerce, my aunts Jeri Mazurek and Mary Jane Galvin, Pi Kemper, Charlotte Heron, Theo Hugs, Betty Close, and Sue Close helped by supplying me with so many of their favorite recipes.

Thanks, too, to my high school and college buddy, Randy Minnich. We have had many great combination outdoor expeditions.

Finally, thanks to the staff at Falcon Press. For the past six years, Falcon has treated me as a valuable member of a great publishing firm. Their cooperation and help have been invaluable.

FOREWORD

Several years ago, Bob Krumm told me some of our high school friends had told him they didn't consider berry picking a manly pursuit—certainly not in the league of hunting and fishing. When I had finished laughing, I wondered who those poor benighted souls had been. Certainly, they had never gone berrying with Bob, and hadn't dodged briars, nettles, and poison ivy to reach a special patch. Nor had they returned hot, scratched, and happily lugging pails which no longer clanked, but squished. I've hunted ducks more comfortably—with less to show for my efforts.

Nor had our untutored schoolmates known how competitive the sport can be. It's unlikely they'd ever gone to a favorite black raspberry patch at 10 a.m., to read in the dry and browning receptacles that the Krumm brothers had been there at 8. Nor wondered how Bob always filled two pails to everyone else's one. (I suspect he was driven by visions of Mrs. Krumm's pies).

No, those who couldn't stoop to pick berries missed a chance to know the outdoor world more fully. For the berrying merged with the hunting, the hiking, the fishing, the birding, into a school after school that gave depth to the regular curriculum. Indeed, it gave, and still gives, more depth to life.

One day, on a fishing trip in upper Michigan, as Bob worked brook trout caught at dawn and blueberries picked in the hot forenoon into a lunch I shall never forget, I realized how deep his knowledge really was. While I visited the outdoors, he could live there. In that school beyond school, though I was older, he was the teacher. And though my love for the wild burned hot, his burned hotter. He vowed he would not live his life behind a desk and, wonder of wonders, he hasn't.

So now I am delighted that he is offering some of the fruits of that life to the rest of us. That he is letting us follow him with clanking pails, down woodland and meadow paths. Further along, we'll find pies and jellies, of course. And if we look and listen closely, we'll find that our hikes become more: pilgrimages, opportunities to touch and taste and know far better the world beyond asphalt that sustains us.

Randy Minnich, alumnus,
Eaton Rapids High School and Albion College

INTRODUCTION

When I reminisce about growing up in Michigan, it seems that wild berries and fruits played an integral part in my first twenty-one years. While my dad liked to fish and hunt, Mother always made sure an outing was not limited to hunting and fishing. If berries were in season, Mom guaranteed that we'd come back with a bucketful.

Mom introduced my sister, brother, and me to blackberries, black raspberries, huckleberries, blueberries, elderberries, and mulberries. Of course, we didn't want to spend our time picking berries, but Mom was firm, so we picked. I never enjoyed picking berries when I could be fishing, but then a pie or a batch of jam or jelly would make the interruption worthwhile.

Somehow or another, my parents taught me to identify the wild berries and fruits in the area. They also taught me to identify useful trees and shrubs. I learned to identify black walnut, shagbark hickory, and butternut trees, not only because they provided tasty nuts to gather, but also because fox squirrels were attracted to them. Thanks to those early lessons, squirrel pie and chocolate chip cookies studded with walnuts or hickory nuts were standard fare at our house.

It wasn't until I was a senior at Albion College that my country experience and formal experience merged. I had dropped my premedicine curriculum and decided to major in biology. One course I took was Field and Systematic Botany, taught by Dr. E. A. Stowell.

Field and Systematic was basically a plant identification course. During the initial lecture, Dr. Stowell stated his hope that, "By the end of the course, you will grow to look at plants as friends. I know it sounds corny, but after you learn to identify them, they will become familiar, like friends."

It was 1966. I was a hopeless romantic at the time. After spending hours in the field with Dr. Stowell, not only did I look on plants as friends, but I became elated if I encountered a rare or especially beautiful specimen.

Dr. Stowell and the quality liberal arts education at Albion College helped me combine the woods skills and knowledge my mother and dad had taught me with scientific methods for a better understanding of the living world around me.

Dr. Stowell has retired from his professorial duties, but he still tromps about the fields and forests of southern Michigan seeking out new friends. With his help and the help of another botany professor at Albion, Dr. Dan Skean, I was able to write this book.

I still get elated when I find a new plant, but I get even more excited when I find a berry patch, for I can take a bit of that patch home and make some of those great treats my mother still concocts.

I have designed this book so you can identify the common edible wild berries and fruits of the Great Lakes region, including Minnesota, Wisconsin, Michigan, Illinois, Indiana, Ohio, and Ontario. I have tried to identify what the plants look like, their preferred habitat, when they flower, and when they ripen.

I wrote the book for use by people with any level of expertise—especially the novice. I have tried to avoid terms not used in daily use, but when plants are in question, it is sometimes impossible to completely avoid botanical terms. Although you will find a complete glossary toward the end of the book, I will introduce the more commonly used terms here. When I mention berry, I am referring to any small, fleshy fruit. All berries are fruits and most berries are edible. A tree has only one stem and usually reaches a height greater than twenty feet. A shrub has multiple stems and usually reaches a height of less than twenty feet.

Your berry picking will benefit from preseason scouting. That way, you'll already have patches in mind when you go berry picking, and you will also have spent more time out-of-doors. The more out-of-doors time you spend, the better you'll appreciate the living world around you. Coupling your outdoor excursions with some other activity—fishing, hiking, bird watching, photography, whatever—you'll discover even more what a complex, living thing this world of ours is.

I hope *The Great Lakes Berry Book* will help you identify and use the wild berries and fruits around you. I hope, too, that in time, when you look at various fruits and berries, you will think of them as dear and useful friends. Happy picking!

HAZARDS

While berry picking is fun, it also poses some hazards. These may come in a variety of forms.

Foremost are poisonous plants. If you have any question about what you are picking, DON'T EAT IT! Color photos accompany listings of edible species in this book. Use these pictures. Also, read the verbal description of the plant. If you aren't sure about a plant, consult another book, or ask a plant expert.

Some plants are benign. They won't kill you, but neither will they taste good, regardless of what you do with them. Other plants also won't kill you, but they'll make you so sick you'll wish you were dead. And some plants give you extreme dermatitis.

In the Great Lakes area, we have to worry about poison ivy and poison sumac. Poison ivy (*Rhus radicans* and *Rhus rydbergii*) occurs in all but the far western seaboard of the United States—California is the only continental state without the plant. All parts of poison ivy contain an oil called urushiol. Contact with urushiol is known to cause blisters and severe dermatitis. If a person inhales smoke from burning poison ivy, or eats the berries, inflammation of the nasal passages, mouth, and throat can occur.

Poison ivy is a vine with glossy compound leaves. Leaves are composed of three leaflets, each pointed at the tip, broadest near the base, and entire, or with a few large teeth. In autumn, the leaves turn hot orange or red before they fall. Poison ivy has white berries, the size of a BB, in clusters of 10 to 20.

Poison sumac (*Rhus vernix*) is even more potent than poison ivy. Fortunately, it only grows in wet, inaccessible swamps. If you're going after highbush blueberries or cranberries, you'd better be on the lookout for this shrub. It varies in height from 6 to 15 feet and has gray bark with smooth, light green twigs.

The sumac leaf is 6 to 14 inches long, pinnately compound with 7 to 13 leaflets, and glossy dark green in color, with entire margins. The small (smaller than a BB) yellowish green "berry" occurs in loose clusters of 10 to 20. The foliage turns brilliant orange at the first frost and really lights up a bog. But beware of this attractive plant: it will make you itch for days.

A number of poisonous plants have berries of which you should beware. Let's just go down the line in alphabetical order.

Baneberry or chinaberry (*Actaea rubra*), an herbaceous perennial, grows in woods and thickets of the boreal forests of the Great Lakes area. The light green leaves are highly divided. The red berries look highly polished and are quite hard, hence their other common name, chinaberry. The poison in baneberry is a glycoside: it can cause quickening of the heartbeat, gastroenteritis, dizziness, diarrhea, and vomiting. Fatalities have been reported. White baneberry (*Actaea pachypoda*) is closely related and has white or red berries with black spots on the ends that look like doll's eyes. This species, too, is poisonous and is more widespread than red baneberry.

Don't eat the leaves or the pits of either black cherry (*Prunus serotina*) or chokecherry (*Prunus virginiana*) because they have hydrocyanic acid in their leaves and in the pits of their fruits. Cooking breaks the hydrocyanic acid into a harmless compound; that is why you can eat chokecherry or black cherry jelly and jam.

Moonseed (*Menispermum canadense*), a vine, resembles wild grape in vine and leaf characteristics and fruit. Moonseed has smooth bark and no tendrils. The grapelike fruits have only one crescent-shaped seed, while wild grapes have numerous egg-shaped seeds. The moonseed fruit, a drupe, is poisonous, producing severe abdominal pain and indigestion. Paralysis and fatalities have been reported.

Both mulberry species (*Morus rubra* and *Morus alba*) contain hallucinogenic compounds in the unripe fruits and in the sap of leaves and stems. The sap can also cause gastric upset.

Woody nightshade (*Solanum dulcamara*) is to be avoided. It has a typical nightshade flower: star-shaped with light purple petals (just like a tomato blossom, but a different color). The fruits look like miniature red Italian tomatoes. Woody nightshade likes moist areas around streams and waste areas. The chemical compound in woody nightshade is solanine, a complex glycoalkaloid that produces two types of poisoning: irritant and nervous. Suffice it to say, solanine can kill you in two ways.

Pokeweed (*Phytolacca americana*), a perennial herbaceous plant, has a purple hue in its stems and leaves. The plant can attain a height of 10 feet, but 4 to 5 feet is more the norm in the Great Lakes area. The stems are upright and branching, and have a disagreeable odor when broken. The glossy, green leaves are alternate and entire. Pokeweed has tight clusters of shiny, deep purple berries that are wider than they are tall. Most of pokeweed's poison is concentrated in the root, but older stems, leaves, and berries contain it as well. You can use the young shoots much as you would asparagus, but you need to boil them in at least two changes of water before eating. Pokeweed can cause stomach burn, vomiting, diarrhea, and disturbed vision. It can also be fatal.

Common buckthorn (*Rhamnus cathartica*), a spiny shrub, has light green, fine-toothed leaves. The plant usually reaches heights of up to 20 feet. The berries, green to black and glossy, occur in the axis between the leaves and the stem. Buckthorn is an escape and grows in waste areas—roadsides, abandoned fields, and the edges of woods—preferring well-drained soils. The berries contain 4 different chemical compounds which can cause severe thirst, vomiting, and violent diarrhea upon ingestion.

Other hazards you might encounter when berry picking would come under an animal heading. The northern part of the Great Lakes area is home to black bears, and berries, most particularly blueberries, are one of their favorite foods. Make it a point to give bears the right-of-way. No bucket of berries is worth getting mauled.

Poisonous snakes inhabit all of the states in the Great Lakes region. The small massasauga rattlesnake is probably the most common in the region, but in more southerly extremes, copperheads are present. Make sure to watch where you place your feet. If you look for snakes, you probably will avoid them.

One of the biggest problems I encounter when berry picking is bees and wasps. If you are allergic to bee or wasp stings, you probably ought to forego berry picking, or carry the appropriate insect antidote syringe with you. The best defense against bees and wasps is to watch for them. You should be able to see and hear them concentrated around their hives. Avoid the hives, for the insects will defend their homes vigorously.

Since many of the wild berries and fruits occur in disturbed areas and roadsides, you must be aware of weed sprays. Many county and state road departments spray right-of-ways. If plants look wilted, brown, or unhealthy, don't pick their berries.

Blackberry
and
Dewberry

Blackberry has always been a favorite of mine, though it causes me approach-avoidance conflict. The plant can easily rip my skin through pants and shirt; but it also produces visions of luscious pies, jelly, jam, and blackberries with milk and sugar.

There were plenty of blackberry thickets near my home town in Michigan. Since blackberries ripened later than black raspberries, I could scout for blackberry hot spots while I picked black raspberries.

While blackberry picking is a fine pursuit in itself, a combination trip used to fill our family's needs best. Some Sundays, while Mom packed a picnic lunch in a wicker basket, we kids would dig worms to fish with, then all trek off crosslots with fishing gear, picnic basket, and high hopes.

Dad would get us set up fishing, then disappear to fish for northern pike while Mom kept an eye on us. We would catch sunfish, creek chubs, shiners, and maybe a carp or two before lunchtime rolled around. Those picnic lunches always seemed tasty.

After fishing some more, we would head home, but not before stopping at one of the woods along the way. Many times, we picked a gallon or so of blackberries before we made it home. Sometimes we would come home with a nice pike Dad had caught and a bucket of berries. There would be good eating at our house for a day or two!

Blackberry not only provided us with tasty berries, but during the winter, the patches provided us with cottontail rabbits. Dad was a good tracker, and he liked to hunt after a fresh snow. He would follow the tracks until he located the rabbit. Many times the cottontail would be sitting in a thick blackberry patch. It took some work to kick the rabbit out of the patch. Often the little creature would make a quick dash to heavy cover and escape, but Dad was successful enough that, during the winter, we had rabbit pie a couple times a month.

My grandmother lived in the hill country near Steubenville, Ohio. My stepgrandfather was quite the wine maker. I remember picking blackberries for that purpose from huge patches around abandoned coal mines. That was a different berry hazard. I've worried about poison ivy, snakes, bees, wasps, and yellow jackets, but open mine shafts were unique to that area.

One of the best blackberry patches I can recall was on some back road near Madison, Wisconsin. My sons and I must have picked a gallon in half an hour.

Some blackberry stems have sparse thorns; others are downright prickly. Regardless, thorns can be a big deterrent, so if you're going to get serious about blackberry picking, you'd better devise a strategy for picking them. Some pickers wear brush pants to keep the thorns from poking into their legs, while others wear heavy leather gloves that enable them to pull stems out of the way as they pick their way in and out of a thicket. One New Zealand berry picker I know employs an extreme method: with hedge clippers, he cuts a path through the brambles. I don't recommend the practice.

Numerous species of blackberry are widespread in the Great Lakes area. All blackberries, as well as raspberries and dewberries, belong to the genus *Rubus*. Since blackberries hybridize readily and taxonomic differences are slight, we will lump them into the following categories: highs, half-highs, and lows. I am most familiar with the highs and lows; it seems that half-highs aren't very productive. The most common high species is *Rubus allegheniensis.* My mother refers to one species of high blackberry as "king" berries. These may be escapes, but they certainly were bigger in length (about 3 cm, or an inch) and diameter (1.5 cm, or 3/4 inch) than the typical species (2 cm, or 1/8 inch in length).

The lows include one species that berry pickers should take note of, namely, dewberry. Dewberry, *Rubus flagellaris,* is a recumbent shrub—that is, it crawls along the ground. Dewberry blossoms and ripens at the same time as the high blackberries.

Most years, you should be able to pick a gallon or two of blackberries— enough for a couple of pies and a batch of jam. Blackberries freeze well, so if you can't put them up quickly, spread them out on trays and put them in the freezer. When they're frozen, place them in quart freezer bags, then back into the freezer until you can get to them.

કે

IDENTIFICATION

Blackberry isn't too fussy about where to grow. It does well in rich, well-drained soils with moderate amounts of sunlight, but you can find it growing in sandy, clay, or gravelly soils along roadsides, fencerows, edges of woods, and waste areas. You might even find it in open woods where blowdowns have occurred, or where trees are widely spaced, allowing some sunlight to reach the forest floor.

Blackberry often has a ribbed, red-hued stem with sickle- shaped thorns. High blackberries range from 3 to 9 feet in height; canes are arching, five-sided, and fluted. Dewberries grow from 4 to 12 inches high. They trail along the ground. The tips of the stems often root.

Both types have leaves that are alternate and palmately compound, normally with five leaflets. On top, leaflets are a light to deep green color, and pale green beneath. Leaflets have pronounced serrations along the margins.

Blackberries and dewberries are biennials, that is, aboveground stems or canes only live two years. The first year, roots produce a long, unbranched cane. The second year, small side branches on the canes produce flowers.

In the southern Great Lakes area, blackberries and dewberries blossom in late May or early June. Further north, they blossom in mid- to late June. The 5-petalled white flowers are arranged in racemelike clusters, 6 to 15 flowers per cluster.

In southern areas, blackberries ripen in late July; in northern areas, throughout August. As it ripens, the thimble-shaped berry changes from green to red to glossy black. When you pick a blackberry or dewberry, the receptacle comes with it, while with thimbleberries and black or red raspberries, the receptacle remains on the stem of the bush—a major difference between raspberries and blackberries.

RECIPES

Blackberries in Framboise

3 pints blackberries
2 cups sugar
2 cups water
1 stick cinnamon, broken
1 tablespoon grated lemon peel
$1/_2$ teaspoon freshly grated nutmeg
$1/_2$ cup Framboise, raspberry brandy

Extract juice from 1 pint blackberries. Measure $1/_2$ cup juice and set aside. Combine sugar, water, cinnamon stick, lemon peel, and nutmeg in large saucepot. Bring to a boil. Reduce heat and simmer 5 minutes. Strain syrup; return to saucepot. Add blackberry juice, remaining blackberries, and raspberry brandy; reheat to a boil. Remove from heat. Pack blackberries into hot jars, leaving half an inch of head space. Pour syrup over blackberries, again leaving half an inch of head space. Remove air bubbles. Adjust caps. Process 10 minutes in boiling water bath. Yield: About three 12-ounce jars.

— From *Ball Blue Book*; submitted by Sue Close, Kiel, Wisconsin

Blackberry Pie

4 cups fresh washed blackberries
I to 1$^1/_2$ cups sugar
2$^1/_2$ tablespoons cornstarch
dash of salt
3 or 4 pats butter
Double pie crust for 9-inch pie

Preheat oven to 425 degrees. Place lower crust in pie pan. Combine sugar, cornstarch, and salt in a small bowl. Combine sugar mixture and berries in a large bowl. Place in pie shell. Dot with butter. Place top crust on and vent. Seal and flute edges. Bake 40 minutes or until pie is golden brown and fruit is tender.

— **Helen Fitzgerald, Lansing, Michigan**

Blackberry Jelly

3$^3/_4$ cups prepared juice (about 2$^1/_2$ quarts fully ripe blackberries)*
4$^1/_2$ cups sugar
I box SURE.JELL® Fruit Pectin
$^1/_2$ teaspoon margarine or butter

*Do not use blackcaps.

Boil jars on rack in large pot filled with water 10 minutes. Place flat lids in saucepan with water. Bring to boil; remove from heat. Let stand in hot water until ready to fill. Drain well.

Thoroughly crush blackberries, one layer at a time. Place 3 layers damp cheesecloth or jelly bag in large bowl. Pour prepared fruit into cheesecloth. Tie chesecloth closed; hang and let drip. When dripping has almost ceased, press gently. Measure 3$^3/_4$ cups into 6- or 8-quart saucepot.

(continued on next page)

Measure sugar and set aside. Mix pectin into juice in saucepot. Add margarine. Place over high heat; Bring to a full rolling boil, stirring constantly. Immediately stir in all sugar. Bring to a full rolling boil and boil 1 minute, stirring constantly. Remove from heat and skim off foam with metal spoon. Ladle quickly into prepared jars, filling to within $^1/_8$ inch of tops. Wipe jar rims and threads. Cover with 2-piece lids. Screw bands tightly.* Invert jars for 5 minutes, then turn upright. After jars cool, check seals.

*Or follow water bath method recommended by USDA.

Makes about 6 cups or 6 (1 cup) jars.

Kraft Foods Inc.

Blackberry Jelly Roll

1 cup flour
4 eggs
1 teaspoon baking powder
1 teaspoon vanilla
dash of salt
1 jar blackberry jelly*
1 cup sugar

Preheat oven to 375 degrees. Measure and sift dry ingredients into bowl. In another bowl, beat eggs and vanilla together, then add the dry ingredients. Beat by hand until all the flour has been added. Pour onto a greased, floured cookie sheet. Bake for 20 minutes. As soon as the roll is done, remove from the oven and peel back edges of roll from sheet. Invert roll onto a dish towel that has been sprinkled with powdered sugar. Cut off crusty edges of roll. Stir up blackberry jelly with a knife; spread jelly onto roll. Roll up quickly, using towel as a guide. Remove towel, and place jelly roll on a rack (seam-side down) to cool.

*About any flavor jelly can be used in jelly roll.

— Emily Krumm, Eaton Rapids, Michigan

Blackberry Jam Cake

1 cup brown sugar
$^3/_4$ cup butter (creamed)
3 eggs
1 $^1/_2$ cup flour
3 tablespoon sour cream
1 teaspoon nutmeg
1 teaspoon cinnamon
1 teaspoon allspice
1 teaspoon baking soda
1 cup wild blackberry jam

Mix ingredients in order listed. Bake in three layers until done (350 degrees for about 30 minutes). Frost with white frosting.

— **Judy Faurot, Sheridan, Wyoming**

Black Cherry

Black cherry brings to mind a piece of land that was only a few hundred yards from our home in Eaton Rapids. The land was what I would term bottomland. It was probably once part of a large swamp or bog, for the soil was deep, black, and lacking any clay or sand. Further west, there were willow tangles and, in times of heavy rain, ephemeral ponds.

In the better drained part of the bottomland grew numerous black cherry trees. Since the trees weren't crowded together, the spaces between them were filled with blackberries and black raspberries. It was one of my favorite berry picking spots, though I must admit I never once sampled any of the plentiful black cherries that appeared there each July. I realize now what a treasure I passed by—but that doesn't prevent me from picking them nowadays.

Black cherry is not only a decent-tasting wild fruit, but its wood is highly prized for furniture making and as cabinet wood. It is nearly as valuable as black walnut, and the finished wood looks much like mahogany.

Unlike chokecherry, its close relative, black cherry or *Prunus serotina* has a fruit that is only slightly bitter-tasting. You can eat black cherries as is, especially if they are very ripe. You can use black cherries for a multitude of recipes, including jelly, jam, syrup, pies, liqueurs, wines, and 'ades. You can use black cherries interchangeably in chokecherry recipes. For those recipes below that require pitted cherries, rather than the juice, remember that many specialty catalogs carry cherry pitters. These can pit the cherry adequately and leave you the pulp.

Your chances of getting a gallon or so of black cherries are about one in four—but when you hit it right, you'll have a gracious plenty.

❧

IDENTIFICATION

Black cherry trees attain heights as great as 100 feet. The trees prefer moist, well-drained, fertile soils with an east or north aspect, or in protected coves. You can find black cherry trees growing in mixed deciduous stands or in pure stands. They are widespread throughout the Great Lakes area, including southern Ontario.

The bark of black cherry is dark gray to nearly black in older trees with scales or plates. The bark of younger trees is smooth, with a reddish brown hue and prominent horizontal lenticels. The branches are also reddish brown, and they droop.

Leaves are alternate, simple, and deciduous, with blunt-toothed margins. They are shiny green on top and light green underneath, often with hairs along the midvein. Leaf shape is elliptical to lance-shaped, from 2 to 5 inches long.

Black cherry blossoms in early May, so at times the cherry crop is frosted out. The creamy white blossoms are approximately $1/4$ inch in diameter, appearing in a cluster (raceme) about 3 inches long. Black cherry ripens in late July, with cherries about the size of large peas. The cherry turns a lustrous black and is edible when ripe.

RECIPES

Black Cherry Jam

Remove stems from black cherries and wash well. Add 1 cup water for every 4 cups cherries. Place over low heat and simmer until fruit is very soft, stirring occasionally. Rub pulp through medium sieve, measure, and add equal amount of sugar. Place over medium heat and stir until sugar is dissolved. Bring to a full rolling boil and cook until mixture sheets on a spoon, or has a temperature of 220 degrees. Seal in hot sterilized jars and process in boiling water bath for 5 minutes. Three cups of pulp will make approximately three half-pints of jam.

– Anonymous, Lewistown Chamber of Commerce, Lewistown, Montana
Montana Chokecherry Festival Recipe Book

Black Cherry Fruit Jelly with Cream

1 1/2 pounds black cherries
6 1/2 cups water
sugar to taste
1/4 cup cornstarch
1/4 cup water
1/2 teaspoon vanilla
slivered almonds
cream

Simmer fruit and water. Sieve to remove seeds. Sweeten pulp to taste with sugar. Bring to a boil. Mix cornstarch with 1/4 cup water; blend into fruit. Cook until thickened. Add vanilla; pour into serving bowl. Cool until thick and set. Decorate with slivered almonds. Spoon into sauce dishes; serve with cream and sugar.

– Donna Davis, Lewistown, Montana
Montana Chokecherry Festival Recipe Book

Sour Cream Black Cherry Pie

4 egg yolks (save the whites)
1 cup pitted black cherries (cooked a few minutes)
$^1/_2$ teaspoon cloves
1 cup sour cream
1 cup sugar
baked pie shell

Cook until thickened in double boiler. Fill pie shell with mixture. Pile meringue on pie, then put in 350-degree oven and bake 15 minutes or until brown.

Alice's never-fail meringue:

scant cup water
6 tablespoon sugar
1 tablespoon cornstarch
egg whites left over from above
$^1/_4$ teaspoon of cream of tartar

Before you start the pie, boil the water, sugar, and cornstarch together; cook until transparent. Cool. Beat the egg whites together with cream of tartar. Pour cooled sugar mixture into beaten egg whites. Beat until stiff, not dry. Put on pie; bake at 350 degrees for 15 minutes or until brown. Meringue will not weep or wilt.

– **Alice Dunn, Lewistown, Montana**
Montana Chokecherry Festival Recipe Book

Black Cherry Cooler

4 cups black cherry juice*
$3^1/_2$ cups sugar
1 (46-ounce) can of orange juice, or 1 (12-ounce) can frozen orange juice, mixed as directed with 3 cups water added

Combine cherry juice and sugar with orange juice. Heat for 15 minutes and cool. Add ice; serve. Makes ten 8-ounce glasses.

*Chokecherry juice can be used interchangeably.

– Ella Hitchcock, Baker, Montana
Montana Chokecherry Festival Recipe Book

Black Cherry Syrup

2 cups black cherry or chokecherry juice
$^1/_2$ cup corn syrup
$2^1/_2$ cups sugar

Bring to a boil. Turn down heat and simmer for 20 minutes. Pour into hot jars and seal.

– Barbara Smith, Billings, Montana
Montana Chokecherry Festival Recipe Book

Black Cherry Frappe

vanilla ice cream
$^1/_4$ cup black cherry syrup*
$^1/_2$ cup ginger ale

For each serving, fill a glass with vanilla ice cream. Add syrup and ginger ale. Stir and serve.

*Can use chokecherry.

Mary Lou Cook, Lewistown, Montana
Montana Chokecherry Festival Recipe Book

Black Raspberry

If I had to choose one berry as my favorite, I would choose black raspberry. Black raspberry has excellent flavor, it is delicious eaten as is, it is widespread and relatively easy to pick, and it has a wide variety of uses.

I have many memories of picking black raspberries outside my home town of Eaton Rapids, Michigan. The flood plain of the Grand River had lush vegetation, with plenty of elms and silver maples. At other places, black cherry trees growing on thick muck deposits provided a bit of shade for the black raspberries.

It seems these shaded, moist areas held the most succulent, tastiest black raspberries to be found. I would wade through high grass, poison ivy, brush, and other obstacles to get at them. Some raspberries were so ripe, they would fall off when I nudged the bush. Those berries I managed to catch before they fell went into my mouth, not my bucket. Boy, were they succulent, and oh so flavorful.

I came back from those berry-picking expeditions with a two-gallon bucketful of scrumptious black raspberries, my hands stained purple. I thought I'd never get rid of the stains, but in a week or so, my hands would look normal again.

The shortest time it ever took me to fill that two-gallon bucket was a little over an hour. That was a berry good time!

Mom made a host of treats from black raspberries, but one was so simple, even I could make it. I came to enjoy black raspberries sprinkled with sugar and covered with milk. Gads, I bet I ate them by the quart. Black raspberries sprinkled over my breakfast cereal were hard to beat, but black raspberries mixed with sugar and sprinkled over vanilla ice cream came close. Of course, black raspberry jam on hot toast could be a complete breakfast, if I wanted to go through half a loaf of bread. Then too, I couldn't turn away from black raspberry pie, with a scoop of vanilla ice cream topping it off.

Black raspberries freeze well, so if you can't make jelly or jam at the moment, or you desire a pie or some other concoction, freeze the berries by spreading them out on a cookie sheet. After they are frozen, place them in quart freezer bags and put them back in the freezer until you can get to them.

Your chances of getting a gallon or two of black raspberries most years, except in severe drought, are excellent. You never know when one of those succulent morsels might spoil in your bucket, so make sure to sample often as you pick. (Just kidding—but a black raspberry treat here and there will help you enjoy your berry tough task.)

&

IDENTIFICATION

Black raspberries are widespread in the Great Lakes region simply because birds love them and the seeds are highly resistant to digestion. Wherever a bird defecates black raspberry seeds, a black raspberry bush is liable to grow.

Black raspberries prefer areas with moderate to full sunlight, so expect to find them in disturbed areas or areas where the vegetation is in the initial stages of a forest, i.e., seedlings and saplings.

As I mentioned, black raspberries thrive in river flood plains mingled with trees that let some light through. However, you can also find them in upland situations, at the edges of cultivated fields, hedgerows, waste areas near cities and towns, and along highways and trails.

Black raspberries have a moderate amount of prickles, but these aren't much of a deterrent, especially compared to blackberries. The canes are light purple in color with a light bloom. Black raspberry canes tend to bend down to the ground. The cane will often take root, creating another clone.

The scientific name for black raspberry is *Rubus occidentalis*. It shares the genus with red raspberry, dewberry, blackberry, and thimbleberry.

Black raspberries blossom in early June and ripen throughout July—in southern areas, in early July, and in northern areas, in late July. My mother used to say the black raspberries were ready when the wheat was ripe. Black raspberries turn from green to bright red to lustrous black. When you pick a black raspberry, the core—the receptacle—stays on the bush, unlike that of a blackberry or dewberry, which remains in the berry.

Black raspberries have a light green, pinnately compound leaf with three or, rarely, five leaflets. The underside of the leaflet is pubescent, giving it a white color.

Bushes seldom exceed 4 feet in height because the stems droop to the ground, but stems may have an overall length of 3 or 4 yards. You may find solitary bushes, or a clump of ten to twenty bushes. Solid stands such as blackberries attain are rare with black raspberries.

RECIPES

Black Raspberry Cake

2 cups flour
$^1/_2$ cup sugar
2 teaspoons baking powder
$^1/_2$ teaspoon salt
$^1/_4$ cup butter
1 egg
1 cup milk
2 cups black raspberries

Mix first five ingredients together like pie crust. Mix in egg and milk. Put mixture in a greased 9 x 13-inch pan. Cover with black raspberries.

Topping:

$^1/_2$ cup butter
1 cup sugar
1 cup flour

Mix together and sprinkle over berries. Bake at 350 degrees for 35 to 45 minutes.

Rose Henckel, Milwaukee, Wisconsin

Black Raspberry Jam

5 cups prepared fruit (about 2 quarts fully ripe black raspberries)
$6^1/_2$ cups (2 pounds 14 ounces) sugar
1 box SURE.JELL® Fruit Pectin

First prepare the fruit. Thoroughly crush one layer at a time, about 2 quarts black raspberries. (Sieve half of the pulp to remove some of the seeds, if desired.) Measure 5 cups into 6- or 8-quart saucepot.

Then make the jam. Measure sugar and set aside. Mix fruit pectin into fruit in saucepot. Place over high heat and stir until mixture comes to a full boil. Immediately add all sugar and stir. Bring to a full rolling boil and boil hard 1 minute, stirring constantly. Remove from heat and skim off foam with metal spoon. Ladle quickly into hot sterilzed jars, filling to within $1/8$ inch of tops. Wipe jar rims and threads. Cover with 2-piece lids.* Screw bands tightly. Invert for 5 minutes, then turn upright. After 1 hour check seals. Makes about 8 (8 fluid ounce) jars.

*Or process in boiling water bath for 5 minutes.

Kraft Foods Inc.

Old-Fashioned Berry Fritters

Prepare fritter batter by mixing together:

- 1 cup flour
- 1 beaten egg
- 1 cup milk (a little less)
- 1 tablespoon baking powder
- $1/2$ to 1 cup black raspberries*

Stir first 4 ingredients together, then add black raspberries—depending on how many you have picked. Drop by tablespoonfuls onto a heated, oiled griddle. Brown on both sides. Sprinkle with powdered sugar. Serve warm.

*About any berry can be used—red raspberry, strawberry, juneberry, blueberry, or huckleberry.

Sue Close, Kiel, Wisconsin

Black Raspberry Pie

I quart black raspberries
I cup sugar
3 tablespoons flour
dash of salt
3 pats of butter
I unbaked pie shell and top crust (see below)

Combine black raspberries, sugar, flour, and salt. Place in pie shell. Add butter; cover with top pie crust. Pierce the crust with a fork in four or five places. Bake at 450 degrees for 10 minutes, then lower the temperature to 350 and bake for 30 more minutes.

Emily Krumm, Eaton Rapids, Michigan

No-Cholesterol Pie Crust

I cup flour
$^1/_2$ teaspoon salt
I stick regular Fleischman's margarine
3 tablespoons ice water (approximately)
I teaspoon vinegar

Combine flour and salt. Cut in margarine. Combine water and vinegar. Add water to flour mixture gradually. Use only enough to make a firm dough. Roll out on a well-floured surface. Make sure to flour rolling pin as well.

Emily Krumm, Eaton Rapids, Michigan

No-Cook Black Raspberry Jam

I quart black raspberries
2 cups sugar
2 tablespoon lemon juice
I pouch liquid pectin

Crush berries with a potato masher. Measure two cups of crushed berries into a large bowl. Add sugar and stir well. Let stand at least 10 minutes. Mix lemon juice and Certo together, then stir into crushed berries—stir for 3 minutes. Pour into clean jars, cover, let set at room temperature for 24 hours. Store in freezer.

Emily Krumm, Eaton Rapids, Michigan

Hobo Cookies

4 cups flour
1 teaspoon vanilla
1 cup shortening
$^1/_2$ teaspoon salt
1 cup butter (or margarine)
3 egg yolks, lightly beaten
1 cup milk
1 8-oz jar black raspberry jelly or jam*

Preheat oven to 350 degrees. Combine flour, shortening, and butter; work together like pie dough. Set aside. Scald milk, cool to lukewarm, dissolve yeast in it, and let rest 10 to 15 minutes. Add vanilla, salt, and egg yolks. Mix thoroughly.

Mix the yeast mixture into the flour mixture. Put sugar on a board as you would flour to roll out pie dough. Roll $^1/_4$- to $^1/_8$- inch thick. Cut into 2-inch squares. Add a heaping teaspoon of jelly or jam to the center of the square. Pinch the corners of the square together. Bake on greased baking sheets 12 to 15 minutes, or until dough just starts to brown on the edges.

*Any of the jams and jellies listed in the book would work for this recipe. Try a variety of them.

Emily Krumm, Eaton Rapids, Michigan

"This is my favorite cookie recipe." Bob Krumm

Blueberry

Blueberry symbolizes wholesomeness, common virtues, and hometown pride. In my mind's eye, the name blueberry conjures up pies, pancakes, syrup, ice cream, and muffins. It seems that about every state in the Great Lakes area and northeast has a place named "Blueberry Hill."

Some towns have blueberry festivals, complete with parades, cooking contests, and other tributes to the common blueberry. This is a highly prized and sought after wild berry. Many berry pickers keep the location of their favorite blueberry patches a close secret; they would almost feel better giving out keys to their front doors than to divulge their blueberry patches.

One of my favorite memories of picking blueberries involves a late summer fishing trip that my friend, Randy Minnich, and I took to northern Michigan when we were young. The trout fishing was phenomenal, so much so that Randy and I were often fished out before lunch—and had time on our hands.

We wandered around the oak and jack pine woods near camp. On one of these hikes, we stumbled onto a huge lowbush blueberry patch. The patch covered several acres. Everywhere we looked were heavily laden bushes.

The berries weren't especially big, but they were tasty, and they were so numerous, it didn't take long to fill the old fishing hat Randy was wearing.

The next day we went back to the patch and picked in earnest. In a short time we had picked enough blueberries to fill a large aluminum washpan. It seemed all we had to do was be thorough, concentrate on each branch, pick every berry (I don't recall seeing any green ones; all were ripe), and not go hip-hopping around the patch. We picked an area that seemed to be about the size of a pool table, but that patch held so many berries that, had we picked all the bushes, we would have been busy for a week and gained a bounteous treasure of twenty bushels.

Randy and I started having blueberries with a little sugar and milk for breakfast. On another occasion we had blueberry pancakes with blueberry syrup and fresh trout. We couldn't eat all the blueberries we picked, so we took most of them home, where our mothers made blueberry pies and jam as well as freezing up some for future use. The berries we brought back were more welcome than the trout. I hooked the largest brown trout I ever caught in Michigan on that excursion, yet when I think of that trip, I recall it as the time I had a trophy-picking of blueberries.

Blueberries freeze well, so if you don't have time to put them up after picking, place them on trays in your freezer—don't wash them first. After the berries freeze, you can put them in quart plastic containers or freezer bags. Then use them a quart at a time to make pies or any other treat you desire.

IDENTIFICATION

Blueberry occurs throughout most of North America. In the Great Lakes area, especially in the northern states, four species of blueberry prevail: sweet blueberry, *Vaccinium angustifolium*; velvet leaf blueberry, *V. myrtilloides*; late blueberry, *V. vacillans*; and highbush blueberry, *V. corymbosum*. Blueberries grace most of northern Minnesota, Wisconsin,

Michigan, and almost all of Ontario. The plants prefer acidic soils, which can be found in boggy areas as well as in upland situations dominated by oak and pine.

Lowbush blueberry reaches a height of 6 inches to 3 feet, depending on the species and nutrients. Highbush varieties reach 4 to 10 feet, and are usually quite prolific since they grow in bogs. Blueberry branches are thin with no thorns or prickles, and are light green, sometimes tinged with red. Leaves are simple, either entire or finely toothed, and vary in color from light to medium green. Leaves measure 0.3 to 1.5 inches long and about 0.2 to 0.3 inches wide, with the widest point near the base.

Plants blossom from May to early June, depending on the species and its latitude. The light pink or white blossoms have a bell, urn, or cylinder shape, and occur singly or in loose clusters of two to ten.

Berries ripen in late July through August, depending on the species. Many blueberry species are powdery blue, but some are dark purple to black. Berries vary in size from a small pea to one and a half times the size of a large pea.

Most notably, blueberries have a "navel"—a scalloped edge with five points. The points are remnants of the flower's calyx, which is positioned above the ovary. When the berry grows, the calyx remains on top of it. The navel is not unique to blueberries; it also adorns apples, pears, juneberries, and haws.

You can't mistake blueberry for any other plant. Just look for the clues: fine woody twigs; the berry's scalloped edge; pea size, plus or minus a little; and blue or black color.

RECIPES

Blueberry Buckle

I egg
$^1/_4$ cup shortening
$^1/_2$ cup sugar
I cup flour
I $^1/_2$ teaspoons baking powder
$^1/_2$ teaspoon salt
$^1/_3$ cup milk
I pint blueberries

Make a batter of first seven ingredients and spread in greased coffee cake pan. Pour blueberries over the batter. Spread the following streusel on top of them:

Streusel:

$^1/_4$ cup butter
$^1/_2$ cup sugar
$^1/_3$ cup flour
$^1/_2$ teaspoon cinnamon

Mix together with a fork until pea-sized nuggets form. Spread over blueberries. Bake at 375 degrees about 45 minutes. Serve with cream or ice cream.

Rose Henckel, Milwaukee, Wisconsin

Sour Cream Blueberry Bread

2 cups unsifted all-purpose flour
1 teaspoon baking soda
$1/2$ teaspoon cinnamon
$1/4$ teaspoon salt
1 cup butter, softened
$3/4$ cup sugar
2 large eggs
1 cup mashed, ripe bananas
   (about 2 medium bananas)
$1/2$ cup sour cream
1 cup blueberries
$1/2$ cup coarsely chopped pecans

Heat oven to 350 degrees. Grease and flour 9 x 5-inch loaf pan. Combine flour, soda, cinnamon, and salt. In a large bowl, with electric mixer, cream butter and sugar until light and fluffy. Add eggs, bananas, and sour cream and beat until blended. Gradually, at low speed, beat in dry ingredients; continue beating just until smooth. Fold in blueberries and pecans. Spoon batter into prepared pan. Bake 1 hour or until golden on top and cake tester inserted in center comes out clean. Let cool completely in pan.

Sue Close, Kiel, Wisconsin

Blueberry Brown Betty

2 cups bread crumbs or graham cracker crumbs
3 tablespoons melted butter
2 cups blueberries
1 tablespoon lemon juice
$1/2$ teaspoon grated lemon rind
$1/4$ cup brown or white sugar
$1/3$ cup hot water

Combine crumbs and butter; stir over low heat until lightly browned. Place a third of crumb mixture in greased baking dish. Arrange 1 cup blueberries in a layer over the crumbs. Sprinkle with half the lemon juice and sugar. Add second layer of crumbs and remaining berries, lemon juice, and sugar. Top with remaining crumbs. Pour water over. Bake in moderately hot oven (375 degrees) 30 to 40 minutes. Serve warm with lemon sauce. Serves 6.

Jeri Mazurek, Woodland, Michigan

Blueberry Sourdough Pancakes
Starter:

I cake or package yeast
2 cups warm water
2 cups flour

Mix well. Place in loosely covered glass or pottery bowl in a warm place (80 degrees) overnight.

Set aside a $^1/_2$ cup or more of starter for the next batch. Add remaining starter (about 2 cups) to:

2 eggs
I teaspoon salt
I tablespoon sugar
I teaspoon soda

Blend and mix with a fork. Add:

2 tablespoons melted shortening

Cook on a hot griddle. After batter is on griddle, sprinkle with blueberries.*

*I have also used wild strawberries, red and black raspberries, juneberries, and blackberries.

Betty Close, Kiel, Wisconsin

Blueberry Pancake Syrup

I cup blueberries*
$1/4$ cup brown sugar
$1/2$ cup water

In a small saucepan, mash blueberries thoroughly. Add sugar and water. Heat to boiling, then simmer for 10 minutes. Serve over your blueberry pancakes.

*Any berries mentioned in the sourdough pancake recipe in this section can also be used to make syrup.

Dot Heggie, Gillette, Wyoming

Crazy Crust Blueberry Pie

Pastry:
I cup all-purpose flour
$3/4$ cup water
2 tablespoons sugar
$2/3$ cup shortening
I teaspoon baking powder
I egg
$1/2$ teaspoon salt

Filling:
I quart blueberries*
I teaspoon cinnamon
I cup sugar
$1/4$ teaspoon nutmeg

Preheat oven to 425 degrees. In small mixer bowl, combine all pastry ingredients. Blend well at lowest speed, then beat 2 minutes at medium speed. Spread batter in 10- or 9-inch deep-dish pie pan.

To prepare filling, combine blueberries, sugar, and spices. Carefully spoon filling into center of pastry batter. Do not stir. Bake 40 to 45 minutes until crust is golden brown.

*This recipe also works well with blackberries, elderberries, black raspberries, or juneberries.

Emily Krumm's modification of a Pillsbury Flour recipe

Chokecherry

Versatile is an apt word for chokecherry: it has many uses, and occupies a wide band across North America, from the Atlantic to the Pacific Coast, from Hudson Bay in northern Ontario to the California-Mexico border.

With all that chokecherry offers, it's amazing that most Great Lakes berry pickers don't bother with it. I have to admit, chokecherry wasn't on my list when I lived in Michigan, but after I moved to Wyoming, where most berries are scarce, I learned what a great fruit it is. Wyoming folks probably spend more time picking chokecherries than any other wild fruit or berry. Why, even in neighboring Montana, with its plentiful huckleberries, one town, Lewistown, hosts a chokecherry festival the first Saturday after Labor Day, complete with cooking contests, parades, chokecherry run, and pit-spitting contest. Now, if a Montana town thinks that much of chokecherry, maybe Great Lakes folks should take notice and give it a try.

When I think of chokecherry, I think of jelly and syrup. I don't usually eat chokecherries as a trail snack because they pack a lot of pucker power, but after they have been ripe for two or three weeks, the potency is lost and they become halfway decent to eat, but don't swallow the pits.

Of all the small berries and fruits, chokecherry is the easiest to gather a lot of. Chokecherry bushes grow to respectable heights, so you don't have to bend over to pick; plus the branches are quite supple, so you can pull high branches down easily. Chokecherry bushes usually occur in groves; once you find a grove, you have plenty of picking. And the fruits occur in clusters, so it's not unusual to pick from ten to twenty large cherries in one pass, and two or three gallons in an hour. When I'm in a good chokecherry patch, I usually don't quit until I have five gallons, and then I have met my needs for the year.

My grown sons, Clint and James, have been fond of chokecherry jelly since they were four or five. They soon learned that chokecherry jelly sure set off their toast, or their peanut butter and jelly sandwich. As youngsters, they picked chokecherries. Now, twenty years later, they still go after them.

When Clint was a teenager, he tried to avoid all the stirring involved in making jelly. He didn't bother to stir in the pectin or sugar. The result was big lumps of sugar crystals scattered throughout the jelly (which I wouldn't discover until I opened a jar in November). The jelly was basically inedible. Nowadays, each August, Clint puts up several batches of properly prepared chokecherry jelly.

Your chances of picking two or more gallons of chokecherries are about three in four. About the only thing that hinders a bountiful chokecherry crop is a late freeze during the blossoming period, or a severe infestation of tent caterpillars.

❧

IDENTIFICATION

Chokecherries, *Prunus virginiana*, are a pioneer species. You will find them along the edges of woods and pastures where they can get lots of sun, usually growing in groves of ten to fifty. Chokecherries prefer more mesic conditions. Plants grow up to 25 feet high, but they average 10 to 12 feet. The bark is brown with prominent lenticels (breathing pores). The dark green simple leaves, 2 to 4 inches long, are alternately arranged with numerous fine teeth.

Small, creamy white, 5-petalled chokecherry flowers open in early May. The flowers and ensuing fruits occur in loose, elongate clusters known as racemes. Chokecherries ripen in August and last well into September. The individual cherry turns from green to wine red to glossy black as it ripens. The cherry averages the size of a large pea. Its pit makes up $^2/_3$ to $^3/_4$ of a cherry's volume, and is round or egg-shaped. The pit and wilted leaves contain hydrocyanic acid: don't eat them. Cooking destroys the acid.

RECIPES

Chokecherry Cordial Pie

   1 9-inch pie shell, cooled
   Chokecherry pudding sauce:
   2 cups chokecherry juice
   3 tablespoons cornstarch
   1 cup sugar
   pinch salt
   $^1/_2$ teaspoon almond extract

Divide juice into $^1/_2$ cup and $1^1/_2$ cup portions. Mix cornstarch and $^1/_2$ cup juice. Heat $1^1/_2$ cups juice with sugar; bring to boil. Add cornstarch/juice mixture. Cook to thicken. Add salt and extract. Cool.

First layer:

   6 oz cream cheese
   $^3/_4$ cup powdered sugar
   $^1/_3$ of a 12-ounce container of whipped topping

Beat together and spread on bottom of pie crust.

Second layer: Pour all but $^1/_2$ cup cooled chokecherry pudding sauce into pie over cream cheese mixture.

(continued on next page)

Third layer: Beat $1/2$ cup reserved chokecherry pudding sauce into $1/2$ of 12-ounce container of whipped topping. Carefully, spread onto pudding sauce. Use a teaspoon for best results. Mound rest of whipped topping around edges or in middle, and add chokecherries for decoration.

Norma Robertson, Lewistown, Montana

Chokecherry Jam

Combine equal amounts of chokecherry pulp (no seeds), unsweetened applesauce, and sugar; stir well. Cook in a large pot until double drops fall from edge of spoon (see "sheet test" in glossary). Pour into glasses and add paraffin. (USDA boiling water bath and screw lids and seals.) This takes a little more time to make than recipes made from pectin, but it's worth it. (I make it with 2 cups pulp, 2 cups applesauce, and 4 cups sugar.)

Mrs. Evelyn Woerner, Fromberg, Montana
Montana Chokecherry Festival Recipe Book

Chokecherry Crown Rolls

$3^3/_4$ to $4^1/_4$ cups unsifted flour
$1/2$ cup sugar
2 teaspoons salt
2 packages yeast
$3/_4$ cup milk
$1/2$ cup water
$1/2$ cup margarine
1 egg (at room temperature)

In large bowl, mix 1 cup flour with the sugar, salt, and undissolved yeast. Combine milk, water, and margarine in saucepan and heat over low heat until liquids are very warm (120 to 130 degrees). Margarine does not need to melt. Gradually add all dry ingredients and beat with electric mixer for 2 minutes at medium speed, scraping bowl occasionally. Add egg and $1/2$ cup flour. Beat at high speed, 2 minutes. Add enough additional flour to make a stiff batter. Cover bowl tightly with foil. Chill 2 hours or overnight. Remove dough from refrigerator; let it warm up and rise slightly, for maybe about $1/2$ hour. Turn dough out onto lightly floured board; divide into 18 pieces. Roll each piece into a rope 15 inches long. Hold one end of each rope in place and wind dough around loosely to form coil. Tuck end firmly underneath. Place on greased baking sheets about 2 inches apart. Cover; let rise until double, about 1 hour. Make indentations about 1 inch wide in center of each coil, pressing to bottom. Fill with chokecherry filling (below). Bake at 400 degrees for 12 to 15 minutes or until done. Remove from pans and cool on wire racks. When cool, drizzle with thin icing.

Filling:

2 cups chokecherries
1 cup chokecherry juice
$^1/_4$ cup sugar
$^1/_4$ cup cornstarch

Pit chokecherries. Blend chokecherry juice, sugar, and cornstarch. Cook, stirring constantly, until thickened and clear. Add chokecherries. Cool.

Donna Ferdinand, Lewistown, Montana
Montana Chokecherry Festival Recipe Book

Chokecherry Jelly

3 cups prepared chokecherry juice (about 4 pounds fully ripe chokecherries and 1 cup water)
$^1/_4$ cup lemon juice
$4^1/_2$ cups sugar
1 package MCP PECTIN
$^1/_4$ teaspoon margarine, butter, or oil

Boil jars on rack in large pan filled with water 10 minutes. Place flat lids in saucepan with water. Bring to boil; remove from heat. Let stand in hot water until ready to fill. Drain well.

Thoroughly crush chokecherries one layer at a time. Place in saucepan. Add water and simmer for 15 minutes. Place in jelly cloth or bag and let drip. When dripping has almost ceased, press gently. Measure 3 cups into 6- or 8-quart saucepot. Add lemon juice.

Measure sugar and set aside. Mix fruit pectin into juice in saucepot. Place over high heat and stir until mixture comes to a full boil. Immediately

(continued on next page)

add all sugar and stir. Bring to a full rolling boil. Add margarine and boil hard exactly 2 minutes, stirrring constantly. Remove from heat and skim off foam with metal spoon. Ladle quickly into prepared jars, filling to within $^1/_8$ inch of tops. Wipe jar rims and threads. Cover with 2-piece lids. Screw bands tightly.* Invert jars for 5 minutes, then turn upright. After 1 hour, check seals.

Makes 6 (1 cup) jars.

*Or follow water bath method recommended by USDA.

Kraft Foods Inc.

Chokecherry Wine

1 gallon ground chokecherries, seeds and all
3 gallons water (well water if possible)
1 cake fresh yeast
8 pounds sugar

Mix everything in a crock. Let stand at least 3 weeks, or until it stops bubbling. Lift pulp out with slotted spoon. Add 4 lbs sugar, or to taste. The less sugar added, the drier the wine. Let stand again for 3 to 4 days before bottling in unsterilized bottles or jars.

Leona Cotton, Great Falls, Montana
Montana Chokecherry Festival Recipe Book

Chokecherry Juice

To can chokecherry juice, cover washed chokecherries with water. (Mashing the chokecherries is optional.) Simmer for $^1/_2$ hour. Drain off juice, pour into sterilized jars, and seal. Place in canner. Cover with warm water and bring to a boil. Boil 10 minutes. Remove jars of juice from water bath and set aside until you have more time to make syrup or jelly.

Kay ?, Lewistown, Montana, *Montana Chokecherry Festival Recipe Book*

Cranberry

It seems so strange to me that I had to move away from the Great Lakes area to discover that cranberry grew in abundance in the numerous bogs of the region. I discovered cranberries in Maine, and found that they didn't have the shrub form I expected. You see, I expected a shrub much like blueberry or huckleberry, because cranberry is in the same family, Ericaceae, or the heath family. These plants tend to be shrubs; but the cranberry species that we most often pick, Vaccinium macrocarpon, which is also the domesticated species, is a low-growing, spindly vine.

It's hard to find a berry if you have the wrong image in your head as you search. I was literally stepping on cranberries until my friend, Jan White, gave me specific directions to a patch. He said, "Drive down the road toward Georgetown, pull over at the black mailbox, cross over the road, walk north along the stone fence until it ends in the bog, then look down."

I followed his directions and, much to my surprise, bright red cranberries were at my feet. (Incidentally, there was water at my feet, too. Make sure to wear rubber knee-high boots when you go cranberry picking). They were as red as the ones you buy in the grocery store, and as big or bigger. Anyway,

Thanksgiving 1987 was marked by cranberry treats made from wild berries.

One thing that a botanist friend, Dr. Sam Ristich, taught me was that the later in the year you pick cranberries, the sweeter they get. If you can find cranberries in December, you can eat them without sugar, particularly if you like sour treats. If a bog doesn't freeze tight, and the snow cover keeps the berries insulated, you are likely to find cranberries in the spring—sometimes as late as when the plant blossoms! In other words, cranberries keep well.

Sharon Henry sent me some early history notes on the cranberry. She said, "When the pilgrims landed in New England, the little crimson berry—now so popular on festive days—was a puny thing about the size of a pea, growing wild in small patches. One gusty day down Cape Cod way, the sea broke through a strip of beach and spread a level three-inch carpet of sand over the best wild cranberry bog existing in those parts. When the people beheld the havoc of the storm the feeling was bitter. But when the smothered plants finally broke through the mantle of disaster, lo and behold, the harvest of berries was multiplied four times in quantity, size, and flavor. Rough words against the elements died on the lips of the pilgrims and those who had spoken harshly of the rampant seas bestirred themselves to bring precious sand to aid in the cultivation of all cranberry bogs throughout the peninsula. So an ill wind brought riches upon its wings."

Henry continues, "Cranberry was originally called *craneberry* because the fruit hung from the multiple little stems or cranes, each berry by itself, and also from the appearance of the delicate little flower with its long, cluster of stamens which bear a real or fancied resemblance to the long bill of the crane. Another source said that they were called craneberries because they were a favorite food of the cranes."

Cranberries store well with normal refrigeration, but if you can't get to them for a long time, freeze them.

IDENTIFICATION

The same cranberry species picked by pilgrims in New England extends in a band to Minnesota. Cranberries grow throughout the northern Great Lakes region, including central Minnesota; northern portions of Illinois, Indiana, and Ohio; most of Ontario; and all of Wisconsin and Michigan. They grow in bogs and wet, acid soils.

Cranberry is a low-growing, vinelike plant seldom exceeding 6 inches in height, with small entire leaves. Though the cranberry is a creeping or prostrate shrub, plants can intertwine and dominate parts of a bog. The leaves, which are evergreen, are leathery with whitish undersides.

Cranberry blossoms in June or early July. The pinkish petals curve back, sort of like a cartoon drawing of an exploded gun barrel. The male and female parts of the flower fuse together to form a "crane's bill."

Cranberries ripen in September and October but, as pointed out earlier, since berries persist through frosts, they can be picked into November or later. Berries vary in size, from half to slightly larger than commercial size. When ripe, they turn bright red.

RECIPES

Cranberry Applesause

2 cups cranberries
2 cups sliced apples
1 cup sugar
³/₄ cup water

Combine ingredients. Cover and cook slowly until the fruit is tender, about 20 minutes. Cool slightly, then beat with a wire whip until fluffy and light. Makes approximately 3 cups sauce.

Sharon Henry, Fort Smith, Montana

Fresh Cran-orange Sauce

$1/_2$ cup orange marmalade spreadable fruit
$1/_2$ cup water
$1/_4$ cup granulated sugar
4 cups fresh or frozen cranberries
zest of 1 small orange, sliced into thin curls*
sugar substitute, to taste (optional)

In a medium saucepan, combine water, spreadable fruit, and sugar; bring to a boil and cook, stirring occasionally, 5 minutes. Add cranberries and bring to just below boiling; reduce heat to low and simmer uncovered, stirring occasionally, until berries pop and sauce thickens, 10 minutes. Remove from heat; stir in orange zest. Let sauce cool to room temperature, about $1/_2$ hour. Taste sauce; add sugar substitute if desired. Cover and chill thoroughly.

*The zest of a citrus fruit is the peel without any pith (white membrane). To remove zest from a citrus fruit, use a zester or vegetable peeler. Scrape off any remaining pith from the peel with a small sharp knife.

Sharon Henry, Fort Smith, Montana

Cranberry Orange Apricot Bread

2 cups all purpose flour
1 cup Quaker oats
1 cup sugar
2 teaspoons baking powder
$1/_2$ teaspoon baking soda
$3/_4$ teaspoon salt
2 eggs
$1/_2$ cup oil
$1/_2$ cup orange juice $1/_3$ cup water
1 tablespoon grated orange peel
$3/_4$ cup chopped cranberries
$1/_2$ cup finely chopped apricots
$1/_2$ cup chopped nuts

Heat oven to 350 degrees. Grease and flour bottom only of 9 x 5-inch loaf pan. Combine flour, oats, sugar, baking powder and soda, and salt, mixing well; set aside. Beat eggs and oil with fork or wire whisk to blend thoroughly; mix in orange juice, water, and orange peel. Add to dry ingredients and mix just until moistened. Stir in remaining ingredients. Bake 1 hour and 15 minutes or until toothpick inserted in center comes out clean. Cool 10 minutes; remove from pan. Cool completely on a rack.

Jeri Mazurek, Woodland, Michigan

Cranberry Cake

$^3/_4$ cup sugar
2 teaspoon baking powder
$^1/_2$ cup shortening
$^2/_3$ cup milk
2 eggs
$^1/_2$ teaspoon vanilla
1$^1/_2$ cups flour
2 cups cranberries
$^1/_2$ teaspoon salt

Cream sugar and shortening; add eggs and flour, salt, vanilla, and baking powder alternately with milk. Fold in cranberries. Pour into greased 7 x 11-inch pan. Bake at 350 degrees for 35 to 40 minutes.

Serve with the following:

1 cup cream
1 cup sugar
$^1/_2$ cup butter

Heat together until sugar dissolves and butter melts. Just before serving, pour warmed sauce over a piece of cake.

Betty Close, Kiel, Wisconsin

Cranberry Pie

1 unbaked 9-inch pie crust

Layer in crust:

1 1/2 cups fresh cranberries
1/3 cup brown sugar
1/3 cup chopped nuts

Mix:

1 egg
1/3 cup margarine
1/2 cup sugar
1/2 cup flour

Spread over cranberries. Bake at 350 degrees for 35 minutes.

Serve with whipped or ice cream.

Judy Heimkes, Brooklyn Park, Minnesota

Cranberry-Orange Relish

3 cups cranberries
2 cups sugar
1 orange

Wash and sort cranberries. Grind up whole orange including peel. Add cranberries and sugar and grind thoroughly. Blend until thoroughly chopped.

This has been a Thanksgiving favorite at the Krumm household for 40 years.

Emily Krumm, Eaton Rapids, Michigan

Hot Spiced Cranberry Drink

1 cup fresh cranberries
$1/_2$ cup water
1 tablespoon honey
$3/_4$ cup pineapple juice
$1/_4$ teaspoon ground allspice
4 whole cloves
1 inch-long cinnamon stick
dash nutmeg

In a medium-sized saucepan combine all ingredients. Bring to a boil over moderate heat. Lower heat and simmer 15 minutes. Pour through strainer and serve hot. Yield: 2 servings.

Kathy Krumm, Jackson, Michigan

Currant

Currants have been a staple berry for centuries. They occur in Eurasia, and some cultivated species have been imported into the United States. My parents had several red currant bushes in the garden. From them, Mom made some of the prettiest jelly I have ever seen. To top it off, the jelly had wonderful flavor.

In recent years, I have found currants a rather easy berry to pick because they can usually be found around some of my favorite haunts: streams and mountainous areas. I have even managed to rope some of my fishing clients into currant picking. During the summer, on camping trips, my sons and I picked lots of currants. We usually saved them until we could get home and put them up.

Many fishermen overlook the fact that they can do something other than fish all day. Usually, fishing isn't that red-hot in the middle of the day, so picking berries can be a welcome diversion. You will find that a more diverse trip results in a more enjoyable, relaxing vacation.

Sometimes you can use the berries you pick in a camp meal. My sons and I often have fresh strawberries, raspberries, or blueberries with our meals, adding them to our pancake batter and making a syrup of crushed berries, water, and brown sugar to put over the pancakes. (Incidentally, you can make an excellent camp syrup from crushed currants, brown sugar, and water.)

Currants have a variety of uses, but they can be dried and used as raisins. You can use them in pies, mincemeat, sauces, syrups, and cordials, so check out your local currants. You might find a vital ingredient for some of your favorite recipes. Your chances of getting two quarts of currants—enough for a batch of jelly—are about one in three.

IDENTIFICATION

Currants and gooseberries belong to the same genus: *Ribes*. Members of *Ribes* share several characteristics: a leaf shaped like that of a maple; a trumpet-shaped flower; and a globular berry with a "pigtail" attached. The pigtail is actually the dried remains of the flower. Members of *Ribes* have an inferior ovary, that is, the ovary is below the flower parts, so that as the fertilized ovary grows, the flower parts are pushed to the top of the berry opposite the stem. It's comforting to know that no member of the genus *Ribes* is poisonous; some members may not taste good, but they won't kill you.

The currants we will consider are black currant, *Ribes americanum*, and swamp red currant, *R. triste*. But other wild and escaped species in the Great Lakes region are palatable and worth picking.

Currants occur throughout the region, preferring rich soils in open woodlands, along streams, and in thickets. Plants blossom in the early spring—late April through May. Most commonly, flowers are yellow or greenish white.

Black currant ripens mid- to late July through August; swamp red currant ripens August through September. Black currant turns a shiny black color, while red swamp currant is a shiny red.

RECIPES

Currant Jelly

Select currants. Remove leaves, but do not stem; wash and drain. In a 6- to 8-quart pot, mash with a potato masher; add $^1/_2$ cup of water for each 2 quarts of fruit. Cook 10 minutes, stirring frequently. Strain through jelly bag. Use $^3/_4$ cup sugar per 1 cup juice. Heat juice; add sugar. Stir until sugar dissolves. Cook, stirring occasionally, until syrup sheets off spoon. Seal in hot, sterilized jars.

Jeri Mazurek, Woodland, Michigan

Currant Pie

$^1/_4$ cup flour
2 tablespoons water
1 cup sugar
1 cup currants
2 egg yolks
1 unbaked pie shell

Mix flour and sugar. Lightly beat egg yolks in water; add to flour and sugar. Mix thoroughly. Mix in currants and place in pie shell. Bake 1 hour at 350 degrees. Cool and cover with meringue. Cook in slow oven (300 degrees) until delicately browned.

Meringue:

2 egg whites
$1^1/_2$ teaspoons lemon juice OR
2 tablespoons powdered sugar
$^1/_4$ teaspoon vanilla

Beat until stiff; spread over pie.

Jean Buchner, Maquoketa, Iowa

Currant Cordial

8 cups currants
4 cups sugar
1 quart vodka

Place in a gallon jar. Stir every few days. Let brew 2 months or more. Strain and bottle.

Charlotte Heron, Missoula, Montana

Spiced Currant Jelly

2$\frac{1}{2}$ pounds currants
2 sticks of cinnamon
$\frac{1}{2}$ tablespoon whole cloves
sugar

Mash the currants and cook until soft. Strain through cheesecloth. Measure juice. Place cinnamon and cloves in a cheesecloth bag. Boil spices in juice for 10 minutes; remove spices. For every cup of juice, add $\frac{3}{4}$ cup sugar. Bring to boil. Boil until jellying stage is reached. Pour into hot, sterilized jars and seal. Process in hot water bath for 10 minutes.

Charlotte Heron, Missoula, Montana

Elderberry

Perhaps I should dedicate my berry books to elderberry; it was elderberry that got me started writing them. My friend, Dot, and I were walking along a railroad track in Freeport, Maine. That spring and summer we enjoyed many walks, encountering numerous wildflowers, birds, vistas, and wild fruits. And each new encounter evoked childhood memories.

One June day we found some blossoming elderberry, and I told Dot about how my folks and us kids would go out in late September or early October to pick elderberry. "We would pick the clusters, put them in sacks, and return home to pick the individual berries off so Mom could make jelly out of them," I said. "We would also scout out black walnuts and shagbark hickory nuts at that time. The Michigan fields seemed to offer so much in the fall."

At this point in my story, Dot interjected, "You ought to write a book."

"No, I don't think I could ever write a book. No one wants to read about berries."

"Yes they do, and they would love to hear your stories about them."

"No way," I said, and left the subject. Well, Dot persisted and prevailed; so here I am writing my third berry book—thanks in part to elderberry, and in part to the persistence of a person who sees possibilities rather than impossibilities.

Elderberry sure does bring a lot of memories. Elderberries are just as much a part of my memories of autumn as picking up walnuts and hickory nuts, enjoying fall colors, raking leaves, and walking along with Dad on pheasant hunts.

Most years you should be able to pick a couple of grocery bags full of elderberry clusters, or cymes. Since elderberries have some toxic compounds, don't eat many unripe berries—you could suffer gastrointestinal upset. See the identification section that follows for more information about that.

IDENTIFICATION

While elderberry (*Sambucus canadensis*) is common throughout the Great Lakes region, it seems to prefer the deep, rich, moister soils along streams, mucklands, and swamps. This 5- to 12-foot-tall shrub grows in a round clump of 10 to 30 pithy stems. The stems can be hollowed out easily; in earlier times, people used them as spiles to tap sugar maple trees. The bark has a light silvery gray tint with prominent lenticels that look somewhat like small warts.

The opposite, dark green, glossy leaves are pinnately compound with five to eleven serrated leaflets. The plant blossoms in June or early July. Creamy white flowers are arranged in rounded clusters, approximately pancake-sized, called cymes. The central flowers of the cyme open first. Elderberry ripens in mid- to late September; the deep purple berries, the size of BBs, have three seeds in them, and are also arranged in pancake-sized clusters.

As mentioned above, elderberries do have some toxic compounds, which means you shouldn't eat many unripe berries. Red elderberry, *Sambucus pubens*, is reported to be toxic; DO NOT EAT RED ELDERBERRIES! Just remember, red elderberry ripens early in the summer, while black elderberry ripens in the fall, around the first frost.

RECIPES

Elderberry-Apple Pie

2 cups elderberries
3 tablespoons quick cooking tapioca
$1\frac{1}{2}$ cups apples (sliced or diced)
$1\frac{1}{2}$ cups sugar
2 tablespoons oleo (margarine)

Combine elderberries, apples, sugar, and tapioca. Put in pastry-lined pan; dot with oleo. Cover with top crust. Bake at 350 degrees for an hour.

Pastry:

2 cups flour
$\frac{3}{4}$ teaspoon salt
$\frac{2}{3}$ cup shortening
4 to 5 tablespoons water

Cut shortening into flour and salt until size of crumbs resembles peas. Add water a little at a time. Divide into two parts.

Jeri Mazurek, Woodland, Michigan

Elderberry Jelly

3 cups prepared juice (about 3 pounds fully ripe elderberries)
$\frac{1}{4}$ cup fresh lemon juice
$4\frac{1}{2}$ cups sugar
1 box SURE.JELL® Fruit Pectin
$\frac{1}{2}$ teaspoon margarine or butter

Boil jars on rack in large pot filled water 10 minutes. Place flat lids in saucepan with water. Bring to boil; remove from heat. Let stand in hot water until ready to fill. Drain well before filling.

Remove stems from elderberries. Place in saucepan and thoroughly crush. Heat gently until juice starts to flow. Reduce heat; cover and simmer 15 minutes.

Place 3 layers damp cheesecloth or jelly bag in large bowl. Pour prepared fruit into cheesecloth. Tie cheesecloth closed; hang and let drip. When dripping has almost ceased, press gently. Measure 3 cups into 6- or 8-quart saucepot. (If needed, add up to $1/2$ cup water for exact measure.) Add lemon juice.

Measure sugar and set aside. Mix pectin into juice in saucepot. Add margarine. Place over high heat; bring to a full rolling boil and boil 1 minute, stirring constantly. Remove from heat; skim off foam with metal spoon. Ladle quickly into prepared jars, filling to within $1/8$ inch of tops. Wipe jar rims and threads. Cover with 2-piece lids. Screw band tighly.* Invert jars for 5 minutes, then turn upright. After jars cool, check seals.

*Or follow water bath method recommended by USDA.

Makes about $4^3/4$ cups or about 5 (1 cup) jars.

Kraft Foods Inc.

Elderberry Wine

Mash 20 pounds of elderberries in a 5-gallon crock. Add 5 quarts boiling water. Cover and let stand 3 days. Strain juice and return it to crock. Stir in $10^1/2$ cups sugar. Let stand until fermentation ceases. (It will stop bubbling and frothing.) Remove scum. Strain and bottle. Let liquid age 1 year.

Charlotte Heron, Missoula, Montana

Gooseberry

For those of us who prefer sour flavors, gooseberry sure hits the spot. Gooseberry pie and gooseberry dumplings are two of my favorite desserts.

I am always amazed at how some people seem to enjoy sour flavors—especially youngsters. When he was seven years old, a young friend of mine, Darrel Nieman, joined his grandfather, his brother, and me on a fishing trip. We stopped at one of my favorite fishing spots adjacent to a small island. Darrel's brother and grandfather started fishing, but when I said I was going to pick gooseberries for our dessert at lunch, Darrel elected to join me.

Darrel tagged along and didn't seem to mind the thorns that the wild roses and gooseberries threw in his path. When we found a laden gooseberry bush, I showed the boy how to pick them. Dutifully, Darrel picked a couple of handfuls, then he asked, "Are these any good to eat?"

I replied that they were quite puckery and sour. Darrel announced, "I'm going to try one."

His face showed not a grimace, but a smile. "I like gooseberries," he stated.

For the next quarter of an hour, Darrel ate many more gooseberries than he put in the bucket. I was worried that eating so many might give him a case of diarrhea, but he said, "I just can't quit eating them!"

Fortunately, I managed to pick enough of the berries for gooseberry dumplings, so we retraced our steps to the boat. I hauled out my Coleman stove and cooking gear, and soon had the dumplings simmering. Needless to say, Darrel made short work of his portion. (By the by, Darrel suffered no ill effects from his gooseberry-eating binge). During the next two days, Darrel became proficient at spotting gooseberry bushes.

Gooseberry and currant are fairly common throughout much of the United States. Around the turn of the century, gooseberries and currants in the Great Lakes area and the Northeast became the target of an eradication program. White pine blister rust was killing a very important timber tree, and it was discovered that members of the genus *Ribes*—that is, gooseberries and currants—served as an intermediate host for the rust. At the time, the northern Great Lakes states, Michigan, Wisconsin, and Minnesota, had valuable stands of white pine and, needless to say, gooseberries and currants aren't as common as they once were, due to the eradication program. But decent areas of gooseberry and currant have grown back in the Great Lakes area.

IDENTIFICATION

Gooseberries are shrubs from 2 to 5 feet in height. The stems can be erect or spreading. The plants usually prefer moist woods, lake shores, and river banks, although some species like upland woods and dry soils. There are several definitive features of the genus *Ribes*: a tubular-shaped blossom; a simple lobed leaf that looks much like a maple leaf; and a globular fruit with a "pigtail" attached. The pigtail is simply the dried floral parts. The ovary is inferior in the blossom; as the ovary grows, the flower parts remain attached.

Not all gooseberry and currants are decent to eat, but at least they are not poisonous. So, though some of the berries may be insipid or inedible, they won't kill you. Gooseberry is usually characterized by prickles or spines. Some species are very prickly; even the cultivar has quite a lot of prickles.

The two most common species in the Great Lakes area seem to be prickly gooseberry, *Ribes cynobasti*, and swamp gooseberry, *R. hirtellum*. Both species prefer moist or rocky woods. The prickly gooseberry has prickles on the berry, which limits its use to strained juice. Swamp gooseberry does not have prickles on the berry.

Other gooseberry species in the Great Lakes area share the same characteristics, so look for the prickly stems, globular berries with a pigtail, and a leaf shaped like that of a maple. The ripe berry is normally light purple to reddish purple.

Gooseberry blossoms early in the spring, usually late April through May. The blossoms are tubular and white, greenish white, or purplish. They occur singly or in clusters of up to 5.

Gooseberry ripens July through August, depending on the latitude. An average-sized gooseberry is about the diameter of an M&M. As stated earlier, ripe gooseberries are purple. Green gooseberries that you see marketed are cultivars; if they were ripe when picked, they would have a light purple shade to them, too. I've always noticed fine, light-colored stripes on ripening gooseberries, much like the seams on a basketball.

RECIPES

Gooseberry Jelly

Wash berries; remove leaves and other trash. Put 3 quarts of gooseberries and 2 quarts of water in a large kettle. Cook over low heat until berries can be mashed easily. Strain with cheesecloth and gently press through a jelly bag. Hang up bag and allow the juice to slowly drip into kettle. When dripping stops, press the bag gently.

Mix:

5 cups gooseberry juice
7 cups sugar
1 box powdered pectin

Wash jars; place lids in boiling water; remove from heat. Measure sugar into bowl and set aside. Place gooseberry juice in a 6- to 8-quart pot; stir in pectin. Stirring constantly, bring to a full rolling boil. Stir in sugar. Stirring constantly, bring to a full rolling boil and boil for 1 minute. Remove from heat. Skim off foam; pour into sterilized jars, wipe off rims, put on flat lids, and screw bands. Invert for 5 minutes. Check seals in 1 hour, or use USDA hot water bath method for 10 minutes.

Judy Heimkes, Brooklyn Park, Minnesota

Gooseberry Trifle

2 cups green gooseberries
1 cup sugar
$1/_3$ cup quick tapioca
2 cups boiling water
1 tablespoon lemon juice

Cook tapioca in boiling water 15 minutes. Cook gooseberries and sugar until soft. Add lemon juice to gooseberries. Combine mixtures. Chill and serve with ice cream.

Rose Henckel, Milwaukee, Wisconsin

Gooseberry Meringue Pie

2 cups gooseberries
$1/8$ teaspoon salt
$1/2$ cup water
1 9-inch baked pie shell
1 cup sugar
2 egg whites
$1/4$ cup flour
4 tablespoons sugar

Cook gooseberries in water until tender. Mix sugar, flour, and salt together; add to gooseberries and cook until thick. Cool; pour into baked shell. Spread with meringue made of egg whites and sugar. Bake in moderate oven (350 degrees) 12 to 15 minutes.

Jeri Mazurek, Woodland, Michigan

Gooseberry Meringue Pie

3 cups fresh gooseberries
$1/2$ cup water
1 cup sugar
$1/8$ teaspoon salt
$1/4$ cup flour
3 well-beaten egg yolks
2 tablespoons melted butter or margarine
$1/4$ teaspoon vanilla flavoring
1 9-inch baked pie shell

Simmer gooseberries in water until done. Drain liquid, but keep it. Mix sugar, salt, and flour together and stir into drained liquid. Add egg yolks and butter or margarine and cook until thick, stirring frequently. Gently fold in gooseberries. When cool, turn mixture into baked pie shell. Top with the following meringue. Bake in a 350-degree oven until nicely browned.

Meringue:

3 egg whites
$^1/_4$ teaspoon cream of tartar (add to egg whites before beating)
6 tablespoons sugar (add gradually)

Beat with electric mixer until whites are stiff.

Betty Kemper, Crete, Nebraska

Gooseberry Slump (Dumplings)

Sauce:

2 cups wild gooseberries*
$1^1/_2$ cups water
$^3/_4$ cup sugar

Dumplings:

1 cup biscuit mix
2 tablespoons sugar
$^1/_2$ teaspoon nutmeg
$^1/_3$ cup milk

In a 2- or 3-quart saucepan or pot, mix gooseberries, water, and sugar. Bring to a boil. Cover, reduce heat, and simmer for 10 minutes. Mix dry ingredients for dumplings. Add milk and mix with a fork. Drop batter by tablespoonfuls atop berry sauce. Cook uncovered at low heat for ten minutes; cover and cook for 10 more minutes. Makes four to six servings.

*Blueberries, huckleberries, or cranberries can also be used.

Wilora Dolezal, Cody, Wyoming.
From the *Billings Gazette Cookbook*, September 25, 1983.

"This is my favorite lunch dessert while camping or on a day fishing trip." Bob Krumm

Grape

Wild grape grew practically everywhere in my childhood habitat of south-central Michigan. Since the plant is a climber, it prospered very well in the large forests in the area, and, for that matter, in most of the Great Lakes region.

I remember the first time my childhood friend, Ted, and I ever smoked, we used a piece of grapevine. Maybe that's why I never took up smoking tobacco: just a couple of draws on that grapevine cigarette cured me.

My folks had an arbor of Concord grapes, so my mother didn't fool around with wild grapes—they were too small and too bitter, and climbed too high for her to bother to pick.

Ironically, it was when I moved to Wyoming, where wild grape is uncommon in most areas, that I became interested in picking it. I found areas along the Big Horn Mountains where wild grape luxuriated. I could pick several gallons of grapes in an hour.

Tasting wild grape jelly for the first time, I wanted more. The jelly has such a strong grape flavor, it's almost like eating one of those candies that concentrate flavor, like sour apple or sour lemon. Wild grape is not sour, it's just a very distinctive flavor that is not mild.

While wild grapes are not fit to eat as is, they do make good jelly and juice. Most years, you should be able to pick several gallons of wild grapes. Be careful not to confuse wild grape with a look-alike that is poisonous, moonseed. See the identification section below for further information.

IDENTIFICATION

Wild grape prefers lowland habitat with well-drained soils. There are at least 5 species of wild grape in the Great Lakes area. Of the 5, *Vitis riparia* is perhaps the most abundant. At least 1 species inhabits every Great Lakes state, with some species occurring even in the region's far northern reaches.

Wild grape climbs on whatever it can wrap its tendrils around. All wild grape species have simple leaves with teeth and, in some cases, lobes. Most species have forked tendrils, one has unforked, that grow opposite the leaves. The tendrils are what enables grapevines to climb so well. Wild grape vine bark is variously described as "shreddy" or "loose."

Wild grapes usually blossom in June, after the danger of frost is passed. The grapes ripen from late August through early October, depending on the species and how far north they occur. More southerly grapes ripen earlier because they blossom earlier, while northerly grapes ripen just before the frost in late September or early October. The number of grapes in a clump varies from eight to twenty or so. Each grape contains 2 to 6 rather large seeds. The fruit's color varies from purple to blue to blue-black, all with a bloom.

It is imperative that you check that grapevine bark is loose and shredded, that there are tendrils, and that the grape contains numerous seeds; that way, you won't confuse wild grape with a poisonous look-alike, moonseed. Moonseed has smooth bark, no tendrils, and a single, crescent-shaped seed.

RECIPES

Wild Grape Jelly

Wash and stem 6 quarts wild grapes. Cover with water, bring to a boil, and simmer 20 minutes. Drain using a sieve with a cheesecloth inside.

> 6 cups wild grape juice
> 1 package powdered pectin
> $7\frac{1}{2}$ cups sugar
> juice of 1 lemon

Add the strained lemon juice to the grape juice; heat to boiling. Add the pectin, and again, bring to a boil. Stir in sugar. Bring to a rolling boil; boil hard for 1 minute, stirring constantly. Remove from heat, skim. Pour into jars and seal. Yield: six half-pints.

"I first made this recipe when I was a sophomore in high school and our home economics teacher took us out to pick wild grapes to make our jelly. It's been one of my favorites ever since."

Stacy M. Stokdyk, Sheboygan, Wisconsin

Venison Jelly

> 8 quarts wild grapes
> $\frac{1}{4}$ cup stick cinnamon
> 1 quart vinegar
> 12 cups sugar
> $\frac{1}{4}$ cup whole cloves

Combine grapes, vinegar, cloves, and stick cinnamon in a kettle. Heat slowly to the boiling point; cook until grapes are soft. Strain, then boil 20 minutes. Heat sugar in oven; add the sugar; boil 5 minutes. Turn into sterilized jars and seal. Process 10 minutes in hot water bath.

Dorothy Patent, Missoula, Montana

Grape and Elderberry Jelly

elderberries
wild grapes
sugar

Wash and stem elderberries and pick them over. Do the same for the grapes. Put the elderberries in a deep pan, add water until you can see it through the top layer of berries, and bring to a boil. Cook until elderberries are soft. Strain through a jelly bag. Do the same for the wild grapes, but use only 2 cups water and $1/4$ cup lemon juice.

For every cup of elderberry juice, add one cup grape juice. For each cup of the mixed juices, add $3/4$ cup of sugar. Bring to a boil, stirring constantly. Test frequently for jellying (see "sheet test" in glossary). When two drops run together off the side of the spoon, pour into sterilized jars and seal. Process in boiling water bath for 10 minutes.

Dorothy Patent, Missoula, Montana

Wild Grape and Citrus Punch

4 cups wild grape juice
4 cups limeade
4 cups orange juice
4 tablespoons sugar

Mix all ingredients together. Chill. A scoop of vanilla ice cream can be added to each cup of punch.

Mary Minnich, Eaton Rapids, Michigan

Hawthorn

Hawthorn always held what looked like edible fruits, but for years, I never bothered to find out. Hawthorn thickets grew on some of the glacial areas near Eaton Rapids; they were always a sure bet for holding cottontail rabbits or even a pheasant or two—but I never heard of anyone picking hawthorn. Maybe the abundance of domestically grown apples and crabapples kept us from harvesting them, or maybe the giant thorns that protect the haws scared us away.

A few years ago, Dot and I were scouring an area for wild plums and grapes, and we could find none. It looked like we would return from our berry picking expedition with empty buckets when I happened to spy a grove of hawthorns, their bright red haws glowing in the midday sun.

We pitched in and in a matter of an hour, we had over a gallon of haws. Surprisingly, the menacing thorns were easy to get around; about the only trouble was picking the higher branches of the 8- to 10-foot shrubs. I would have to grab onto a branch and pull it down while Dot picked it clean. With this team-picking method, we got plenty of haws.

Dot made up a batch of hawthorn jelly from those haws, and it had to be the prettiest jelly I've ever seen: a deep, clear red jelly that surpassed the best crabapple jelly in looks and taste.

While haws make great jellies, teas, and jams, you can also eat them as is for survival food. Haws persist throughout the winter. I find that when I am out cross-country skiing or late-season bird hunting, a few haws help when I need something to munch on, though I spit out the seeds and most of the pulp after I've sucked out the juice.

Haws provide a vital winter food source for birds and mammals. Cedar waxwings feed on them, as do ruffed grouse, pheasants, and turkeys. The low-lying branches provide cover for cottontail rabbits, while deer browse on the branches' tips. Porcupines manage to work around the thorns to eat the bark on some of the larger branches.

Most years you should be able to pick a gallon or two of haws. If you aren't far from your car, you might consider packing along a small stepladder to enable you to pick more easily.

IDENTIFICATION

Several species of hawthorns occur in the Great Lakes region including fleshy hawthorn, *Crataegus succulenta*; entangled hawthorn, *C. intricata*; cockspur hawthorn, *C. crus-galli*; dotted hawthorn, *C. punctata*; round-leaved hawthorn, *C. chrysocarpa*; fanleaf hawthorn, *C. flabellata*; downy hawthorn, *C. mollis*; scarlet hawthorn, *C. coccinea*; and frosted hawthorn, *C. pruinosa.*

Hawthorns reach a height of 8 to 25 feet. They usually have spreading spiny branches with 2- to 3-inch-long unbranched thorns and rounded, dark brown winter buds. The leaves are deciduous, alternate, simple, toothed, and shallowly to deeply lobed. Flowers are bisexual, produced singly or in clusters called corymbs. The flowers have a 5-part calyx and 5 petals. Flowers are white or, infrequently, pink or red. The fruits, which resemble miniature apples, are usually red but sometimes yellow, green, or almost black.

That pretty well sums up hawthorns, but I might emphasize that the 2- to 3-inch-long unbranched thorn and the applelike fruits (called haws) are pretty indicative. Part of the blossom is retained opposite to where the stem attaches. That part of the blossom, called the calyx, is star-shaped. If you look at a typical apple (e.g., Delicious or Granny Smith), you will see the same type of arrangement of the calyx; it is very common among members of the rose family.

Hawthorns are a pioneer species; they invade abandoned crop lands and disturbed areas along roadsides, as well as the margins of woodlands. Hawthorns blossom late May to mid-June, and ripen in September or early October. Haws persist through winter.

RECIPES

Hawthorn Jelly

Wash about 1 gallon of haws and place in a 6- to 8-quart pot. Barely cover with water; simmer until soft. Strain through a jelly bag. Measure juice; place in pot; bring to a rolling boil; skim and add $3/4$ cup sugar for each cup of juice. Stir until sugar dissolves; cook until syrup sheets off spoon. Seal in hot, sterilized jars.

Jeri Mazurek, Woodland, Michigan

Thornapple Jam

5 quarts thornapples (haws), reduced to 7 cups pulp
$1/4$ cup lemon juice
1 cup apple juice
$1/2$ cup water
1 package powdered pectin
$1 1/3$ cups honey

Cook thornapples in 2 cups water until they begin to pop (about 20 minutes). Press the cooked apples through a sieve or food mill. To a large saucepan or kettle, add pulp, lemon juice, apple juice, and water. Slowly stir in pectin until dissolved. Add honey. Bring to rolling boil; boil 1 minute. Remove from heat. Pour into sterilized jars, seal, and process 10 minutes in boiling water bath. Yield: eight 8-oz jars.

Kathy Buchner, Jackson, Wyoming

Hawthorn Jelly

3 cups hawthorn juice
$1/4$ cup lemon juice
1 package powdered pectin
$4 1/2$ cups sugar

Wash and sort approximately 5 pounds of hawthorns. Place in pot, crush with potato masher, add 1 cup water. Bring to a boil. Reduce heat and simmer uncovered for 15 minutes. Extract juice.

Place 3 cups hawthorn juice, lemon juice, and pectin in a 6- or 8-quart saucepan. Stir thoroughly. Bring to a full rolling boil, stirring constantly. Stir in sugar. Bring to a full rolling boil, stirring constantly. Boil for 2 minutes. Skim. Pour into hot, sterilized jars. Seal. Process in boiling water bath for 10 minutes.

Charlotte Heron, Missoula, Montana

Highbush Cranberry

Highbush cranberry isn't a real cranberry. It doesn't belong to the genus Vaccinium; *rather, it is a member of the honeysuckle family,* Caprifoliaceae, *and the genus* Viburnum.

Whenever I think of highbush cranberry, I see a calendar photo—a winter setting: a cedar waxwing sitting among clusters of highbush cranberries, one of the shiny red berries in its beak.

Highbush cranberry may be substituted for regular cranberries. Since highbush cranberries are sour and tart, it is best to pick them late in the fall or in the winter to give them a chance to sweeten up a bit. When you are using highbush cranberries in a recipe, remember to strain out the seeds.

Most years, your chances of finding two quarts of highbush cranberries are excellent; four out of five times you'll find plenty. You can pick highbush cranberry throughout the fall and winter, because birds usually wait until late winter before they concentrate on it.

A cautionary note is in order. A shrub of the same genus, guelder rose (*V. opulus*), resembles highbush cranberry. It isn't poisonous, but its fruit is very bitter. So make sure you taste what you pick before you go and put it in a recipe. See the identification section.

ೈ✤

IDENTIFICATION

Highbush cranberry, *Viburnum trilobum*, requires a little more water than most shrubs (it's a lot like a willow in this respect) so you will find it growing along streams and in moist thickets and woodlands—the same places where I expect to find elm trees and silver and red maples. In the Great Lakes area, look for highbush cranberry in most of Minnesota, Wisconsin, Michigan, and Ontario, and in the northern portions of Illinois, Indiana, and Ohio.

Highbush cranberry reaches heights of 6 to 16 feet. The erect to spreading shrub has several ash gray main stems, while young twigs are reddish brown. The leaves are oppositely arranged, simple, three-lobed, and coarsely toothed, with palmate venation. The buds are flattened, and look like small brown mittens. There are no buds at the end of twigs.

Highbush cranberry blossoms in June or July, depending on the latitude. The blossoms are in a floral arrangement called a cyme—a flat arrangement in which the central flowers unfold first. The cyme is 2 to 4 inches across. The petals are creamy white.

Highbush cranberry ripens in September or October. Scarlet red leaves accent the clusters of red egg-shaped berries. Cymes hold ten or more berries. The translucent berries are a $1/_4$ inch to $1/_2$ inch in diameter, and contain a single flattened seed.

Guelder rose, an introduced, escaped Eurasian variety, closely resembles highbush cranberry. While not poisonous, its fruits are so bitter it is inedible. Make sure to taste what you are picking first.

RECIPES

Highbush Cranberry Syrup

2 cups highbush cranberry juice
2 cups sugar
2 cups white corn syrup

Extract juice as for jelly. Heat juice to boiling, add sugar, stir until dissolved. Add corn syrup, heat to boiling, and simmer about 5 minutes. Seal in hot, sterilized jars. Delicious on ice cream, pancakes, or waffles.

Sharon Henry, Fort Smith, Montana

Highbush Cranberry Milk Shake

1 cup milk
$1/4$ cup plus 2 tablespoons highbush cranberry syrup* (above)

Combine milk and syrup in a blender and process until smooth.

Yield: 1 serving.

*You can make berry milk shakes out of any other berries you care to make syrup from.

Sharon Henry, Fort Smith, Montana

Highbush Cranberry Jelly

Crush 2 quarts ripe berries. Add $1/2$ cup water. Simmer 10 minutes. Strain out juice. To 4 cups juice add 1 package powdered pectin. Stir in thoroughly. Heat to boiling and stir for 1 minute. Add 6 cups sugar. Boil hard for 1 minute stirring constantly. Skim and pour into hot, sterilized jars. Seal. Process in hot water bath 10 minutes.

Charlotte Heron, Missoula, Montana

Highbush Cranberry Catsup

1 pound chopped onions
4 pounds highbush cranberries
2 cups water

Combine and cook over moderate heat until soft. Put through a sieve. Add:

2 cups vinegar
2 cups white sugar
2 cups dark brown sugar
1 tablespoon ground cloves
1 tablespoon cinnamon
1 tablespoon allspice
1 tablespoon salt
1 teaspoon pepper

Boil until thick. Pour into hot sterilized jars. Seal. Process in hot water bath 15 minutes.

Very good with wild meat!

Charlotte Heron, Missoula, Montana

Huckleberry

Huckleberry evokes visions of Huckleberry Finn, unhurried times, and tasty treats. Though huckleberry is as American as apple pie, not many berry pickers actually go after huckleberries. Most people mistakenly pick blueberries and think they are huckleberries. Though huckleberries and blueberries are in the same family, Ericaceae, they are not in the same genus.

Huckleberry belongs to the genus *Gaylussacia*, while blueberry is of the genus *Vaccinium*. The major difference between the two is that huckleberry has ten small chewy seeds, while blueberry has numerous minute seeds.

Another name for huckleberry is crackerberry. That's not because it tastes like crackers, but because its seeds crack between your teeth.

I find it interesting that Montana and Wyoming berry pickers talk about picking huckleberries and what wonderful huckleberry patches they have, while in truth, no real huckleberries grow west of Minnesota.

While botanical differences between the two genera exist, most people would agree that huckleberries and blueberries taste about the same. Some might argue that blueberries have just a tad more flavor and sweetness, but it's a small difference.

Huckleberries are edible as is, and since they grow on shrubs that range from 2 to 4 feet tall, you don't have to shed your backpack to pick a handful.

You can make a variety of concoctions with huckleberry, from jams, jellies, and syrups, to pies and tarts. Why, you can even toss some into your pancake batter to create huckleberry pancakes.

You can store huckleberries in the refrigerator for a week or so, or freeze them as you would blueberries. Your chances of getting 2 quarts are three out of four.

ॐ

IDENTIFICATION

Huckleberries are more widespread in the Great Lakes area than blueberries. You can find huckleberries in all the states that comprise the Great Lakes region as well as Ontario and Quebec. The most common huckleberry in the region is black huckleberry, *Gaylussacia baccata*; its habitat is rocky woodlands and boggy areas.

Huckleberry is a shrub that has many branches. As stated earlier, its height ranges from 2 to 4 feet. The bark is uniformly dark to the base of the shrub. Black huckleberry grows in fairly dense clones and often forms monotypic stands in the understory of forests. The oval or oblong leaves are alternate, 1 to $1^1/_2$ inches long, with resinous dots underneath. If you place the underside of the leaf to the back of your hand, you'll be left with a bright yellow mark. Blueberry lacks these resinous dots.

Huckleberry's blossom resembles that of blueberry. The reddish hued, urn- or bell-shaped blossom appears in May or early June, in small clusters.

Huckleberries mature in July. Black huckleberry has a deep purple to black color with a bloom. A huckleberry is approximately the size of a pea.

RECIPES

Steamed Huckleberry Pudding

1 1/2 cups huckleberries
2 1/2 cups flour
1 egg, lightly beaten
3 tablespoons butter
1 tablespoon sugar
2 heaping teaspoons baking powder
milk

Mix all dry ingredients together. Cut butter into dry ingredients. Add the huckleberries and, finally, the egg, with just enough milk to make the mixture stick together. Stir carefully with a knife. Place in a double boiler and steam 1 hour. Serve hot with sugar and cream.

From *Reliable Recipes from Reliable People*, by the ladies of the Third Division, Presbyterian Church, PawPaw, Michigan. Modified by Jean Buchner, Maquoketa, Iowa

Huckleberry Cookies

2 cups sugar
1 cup butter
5 eggs
1 cup milk
1 teaspoon vanilla
3 cups flour
1 teaspoon cream of tartar
1 teaspoon soda
pinch salt
1 pound huckleberries

Blend sugar and butter together until creamy. Add eggs, milk, and vanilla and mix until blended. Mix dry ingredients together and combine with wet ingredients. Fold in huckleberries.

Drop on hot griddle and bake in quick oven: preheat griddle in 400-degree oven, drop cookie dough by the tablespoonful onto griddle, and bake in oven (400 degrees) 10-12 minutes.

Reliable Recipes from Reliable People, **by the ladies of the Third Division, Presbyterian Church, PawPaw, Michigan. Modified by Jean Buchner, Maquoketa, Iowa**

A Peach of a Huckleberry Cobbler

1 tablespoon cornstarch
1 cup fresh huckleberries*
$1/4$ cup brown sugar
1 tablespoon butter or margarine
$1/2$ cup cold water
1 tablespoon lemon juice
2 cups sliced fresh peaches

Mix first three ingredients (left column); add fruits. Cook and stir till mixture thickens. Add butter and lemon juice. Pour into $8^1/4$ x $1^3/4$-inch round pyrex cake dish.

Cobbler crust:

1 cup all-purpose flour
$1/2$ cup granulated sugar
$1^1/2$ teaspons baking powder
pinch of salt
$1/2$ cup milk
$1/4$ cup soft butter or margarine
2 tablespoons sugar
$1/4$ teaspoon nutmeg

Sift together first four ingredients. Add milk and butter or margarine. Beat smooth. Spread over fruit. Mix 2 tablespoons sugar and the nutmeg to make nutmeg topping; sprinkle over top.

Bake cobbler in moderate oven (350 degrees) 30 minutes or until done. Serve warm with cream or milk. Makes 6 servings.

* Blueberries can be substituted.

Emily Krumm, Eaton Rapids, Michigan

Huckleberry Pie

4 cups fresh huckleberries*
$^1/_4$ cup tapioca
$^1/_3$ cup honey
2 tablespoons melted butter
$^1/_2$ teaspoon ground cinnamon (optional)
$^1/_4$ teaspoon ground nutmeg (optional)
I double pie crust

Preheat oven to 450 degrees Roll out one pie crust on a well-floured surface and place in a 9-inch pie pan. In a medium-sized bowl, mix huckleberries, tapioca, honey, butter, cinnamon, and nutmeg. Pour mixture into pie pan. Roll out remaining pie crust and place over pie filling. Seal, trim, and crimp edges. Prick top crust with a fork in four or five places to vent. Bake 15 minutes at 450 degrees; lower temperature to 350 degrees for 45 to 60 minutes more. Remove from oven and cool to room temperature before serving.

* Most all other berries can be used: blackberries, blueberries, cranberries, currants, elderberries, gooseberries, juneberries, raspberries, or strawberries. For the less tart berries, reduce honey to $^1/_4$ cup.

Kathy Krumm, Jackson, Michigan

Huckleberry-Raspberry Jam

2 cups huckleberries
4 cups crushed raspberries (red or black)
7 cups sugar
I bottle liquid fruit pectin

Crush berries. Measure 4 cups (if necessary, add water to come up to full amount). Add sugar, mix well, and bring to full rolling boil. Boil hard for I minute, stirring constantly. Remove from heat; stir in pectin; skim. Ladle into hot sterilized jars, seal. Makes ten half-pints.

Huckleberry Recipes, **Swan Lake Women's Club, Swan Lake, Montana**

Huckleberry Pancakes

1 egg
1 cup milk
1 tablespoon baking powder
2 tablespoons salad oil
2 tablespoons sugar
1 cup flour
$^1/_2$ cup huckleberries*
$^1/_2$ teaspoon salt
more milk, as needed

Beat egg; add milk and oil. Add dry ingredients, mix well, and add huckleberries. Add enough milk until batter is the consistency of thick cream. Bake on hot griddle.

* You can substitute blueberries, raspberries, strawberries, juneberries, or blackberries.

Huckleberry Recipes, Swan Lake Women's Club, Swan Lake, Montana

Juneberry

Juneberry has a number of common names, including serviceberry, shadbush, Saskatoon berry, Indian pear, Indian-cherry, and swamp sugar pear. In the western United States, most people use the term serviceberry, pronouncing it "sarvisberry."

In the eastern United States, people often refer to the plant as juneberry or shadbush. It's called juneberry because it ripens in late June, and shadbush because it blossoms at about the same time that shad move into the rivers to spawn. Names with Indian connotations refer to Indian peoples' extensive use of the berry, especially in pemmican.

Regardless of what it's called, juneberry is a great berry to pick. It makes a delicious trail snack that I enjoy on day hikes to my favorite fishing spots. If the birds haven't already gotten to my favorite clumps of juneberry, I can often pick enough for a pie or jelly—that is, about 2 quarts.

When I was trying to collect photos of juneberry for this book, I realized how popular juneberry is. I found a large juneberry bush, about 12 feet high, covered with light green and fuchsia-colored berries, but not deep-purple ripe berries. I decided I'd come back in four days and photograph the ripe and ripening berries.

On the appointed day, I arrived at the juneberry bush to find that there were a few ripe deep-purple berries and quite a few fuchsia-colored berries. I thought I might as well photograph a clump of berries with ripe, ripening, and green juneberry. I searched the bush. While there were several such clumps at the top of the bush, there were only a couple at a photographable level with all three stages of berries. Finally, I found one I could photograph.

As I was taking the pictures, I detected something in the bush above me. A cedar waxwing was helping himself to the ripe juneberries on the upper part of the bush. He circled the bush and took a half-dozen juneberries before flying off.

I came to the conclusion that if I was going to pick any juneberries, I had better find a bigger patch so both the local birds and I could have our fill. Later, I did find about a five-acre patch of juneberry growing among some quaking aspen. Dot and I picked over 3 gallons in a couple of hours.

My problem with the birds was not unusual for most of the wild berries and fruits that you pick. In the scheme of things, these berries and fruits were designed to appeal to animals so that animals would eat them. The seeds of most berries and fruits are resistant to digestion—yet the seed coat needs to pass through the digestion process. This breaks down the seed coat enough for the plant embryo to emerge.

I have competed with black bears and raccoons for such wild treats as chokecherries, wild plums, and buffaloberries. Bears are notable for their fondness for juneberries, chokecherries, blueberries, and huckleberries. A lot of birds will go after juneberries, gooseberries, currants, wild raspberries, black raspberries, and wild strawberries.

What I'm trying to get at is that you should have several patches in mind when you go after almost any kind of wild berry, because if the birds and mammals don't get them, another human berry picker might, or a late frost or disease might destroy the crop. Of all the wild berries I go after, I would rank juneberry as one of the chanciest for consistently yielding a good picking.

ॐ

IDENTIFICATION

Juneberry belongs to the genus *Amelanchier.* Four major species grow in the Great Lakes area, though one species, downy juneberry, *Amelanchier arborea,* predominates. Other species are Saskatoon serviceberry, *A. alnifolia;* roundleaf juneberry, *A. sanguinea;* and inland juneberry, *A. interior.*

Juneberry is another member of the rose family. The species can be shrubs or small trees. Most juneberry species range in height from 3 to 20 feet, though some grow as high as 40 feet. Juneberry bark is thin and relatively smooth, with minor ridges. Bark color ranges from light gray to brown tinged with red.

The leaves are arranged alternately. Downy serviceberry has leaves 2 to 3 inches long with fine teeth. The leaf has a slight point, and is broadest near the middle. The upper side of the leaf is deep green, while the underside is lighter green with numerous fine hairs.

Juneberry flowers in late April and into May depending on the latitude—the further north, the later it flowers. Suffice it to say, juneberry will flower at nearly the same time that fiddlehead ferns are emerging, and before dogwood blossoms; it will probably be the only shrub blossoming in the woods. Juneberry blossoms are arranged in racemes, three to fifteen in a cluster. Each star-shaped blossom has 5 white petals. The flower clusters nod over, reminding me of a bunch of small pale Japanese lanterns in the spring woods.

As in many members of the rose family, juneberry retains the calyx, so there is a star-shaped growth on the berry opposite the stem. Juneberry ripens from late June to mid-August. It turns from light green to fuchsia to deep purple. The berries are about twice the size of a large pea, and are shaped like apples.

Juneberries keep in the refrigerator for a week or so. If you want to have some for muffins or pies in the winter, I suggest that you either can or freeze them. If you freeze them, don't wash them first. Spread them out on trays and put them in the freezer. After they have frozen, gather the berries into quart-sized freezer bags and return them to the freezer. That way, when you need a handful of berries, you can remove them from the bag without their sticking together. They will keep well for up to eight months in the freezer.

RECIPES

Juneberry Pie

I quart juneberries
I cup sugar
3 tablespoons flour
2 tablespoons lemon juice
Dash salt
3 to 4 pats butter
I complete pie crust*

Combine juneberries, sugar, flour, lemon juice, and salt. Place in unbaked pie shell. Dot with butter, put top crust in place, pierce crust with fork in four spots. Bake at 450 degrees for 10 minutes; lower oven to 350 degrees and bake for 30 more minutes.

*See no-cholesterol pie crust recipe in Black Raspberry chapter, or Evelyn's best pie crust recipe in this chapter.

Emily Krumm, Eaton Rapids, Michigan

Juneberry Muffins

$^3/_4$ cup juneberries*
2 cups flour
$^1/_4$ teaspoon salt
3 teaspoons baking powder
2 tablespoons sugar
$^3/_4$ cup milk
I egg, well beaten
3 tablespoons butter or margarine, melted

(continued on next page)

Preheat oven to 400 degrees. Generously grease a 12-muffin pan. Rinse juneberries and dry on towels. Combine flour, salt, baking powder, and sugar. Sift together into a large bowl. Combine milk, egg, and butter. Add to flour mixture. Stir until just well-moistened. Fold in juneberries. Spoon into prepared muffin pan; fill cups two-thirds full. Bake 25 minutes, or until muffins are lightly browned and toothpick or knife comes out clean. Cool muffins in pan 5 minutes before removing. Muffins are best when served hot out of the pan!

*You can substitute blueberries, strawberries, huckleberries, or cranberries.

Evelyn Hejde, Aladdin, Wyoming

Juneberry Jelly

Start with $3\frac{1}{2}$ pounds juneberries. Place in saucepan and crush. Add $\frac{1}{2}$ cup water. Bring to boil. Reduce heat; cover and simmer 10 minutes. Pour into jelly bag which is in a large bowl. Tie bag closed and hang up. Let drip into bowl until it stops. Press gently. Measure juice. If necessary, add up to $\frac{1}{2}$ cup water for exact measure.

> $3\frac{1}{2}$ cups juice
> 1 package powdered
> 4 cups sugar

Measure sugar into separate bowl; set aside. Stir pectin into juice. Add $\frac{1}{2}$ teaspoon margarine or butter to reduce foam. Bring mixture to a full rolling boil on high heat, stirring constantly. Add sugar. Stirring constantly, bring mixture to a full rolling boil for exactly 1 minute. Remove from heat. Skim off foam. Pour into hot, sterilized jars; seal and place in hot water bath for 10 minutes.

Charlotte Heron, Missoula, Montana

Nontraditional Pemmican

3 cups jerky, ground into powder (can use a blender)
3 cups fresh or frozen juneberries
3 cups raw sunflower seeds
2 tablespoons raw honey
$3/4$ cup peanut butter

Mix ground jerky with berries and seeds. Mix honey and peanut butter over low heat until well blended. Blend all ingredients together. Press into balls the size of a golf ball. Great for a snack on the hiking trail.

Theo Hugs, Fort Smith, Montana

Juneberry Coffee Cake

$1/2$ to $3/4$ recipe of a yeast-type coffee cake dough
1 cup juneberries
1 cup milk mixed with 2 tablespoons dry milk
2 eggs
$1/2$ cup brown sugar, or $3/8$ cup honey
dash nutmeg
$1/2$ cup fine bread crumbs
$1/2$ cup chopped walnuts or sunflower seeds
$3/8$ cup brown sugar

Make coffee cake dough as usual. Roll to fit into a 9-inch pie pan. Prick dough with fork to prevent excessive rising. Sprinkle juneberries on dough. Beat milk, eggs, brown sugar or honey, and nutmeg, and pour over berries. Mix last three ingredients and sprinkle over custard mixture. Bake at 375 degrees for 25 minutes.

Nora Fighter, Fort Smith, Montana

Pemmican

5 cups pounded jerky
4 cups crushed chokecherries or serviceberries (juneberries)
$^1/_4$ cup melted marrow, tallow, or lard

Mix first two ingredients well. Gradually add hot fat until mixture is moist enough to stick together. Form into small balls (golfball-size). Add honey or sugar to sweeten.

Theo Hugs, Fort Smith, Montana

Evelyn's Best Pie Crust Recipe

4 cups flour
$^1/_2$ cup water
1 tablespoon sugar
1 tablespoon vinegar
1 teaspoon salt
1 egg, beaten
$1^3/_4$ cups shortening

Combine dry ingredients. Cut in shortening. Mix together water, vinegar, and beaten egg; add to shortening mixture. Stir together. Chill for at least 15 minutes. Roll out on a well-floured surface. Yield: enough crust for two 2-crust pies.

Evelyn Hejde, Aladdin, Wyoming

"I've used this recipe for over ten years and it hasn't failed me once!"

-Bob Krumm

Mulberry

One mulberry tree can produce gallons of blackberrylike fruit, yet few people take advantage of mulberries. Mulberry doesn't have as rich a flavor as blackberry, but it's tasty nonetheless, and makes great pies and jelly.

My most outstanding memory of mulberry is, as a twelve-year-old, climbing a big mulberry tree because none of the berries were close enough to the ground for me to reach. The trunk had few branches for me to climb on, but I managed to get my legs around the tree and shinny up. I got on a big branch that angled upward at no more than a 15-degree slope, so I pulled myself along until I reached a spot where there were some branches laden with ripe mulberries.

I leaned out and picked a handful. I quickly ate them and leaned out further for some more. I couldn't quite reach them, and then I lost my balance. I grabbed for the trunk but my hold was precarious. I slid all the way down the trunk at quite a clip. I wasn't wearing a shirt, so the rough bark scraped my belly. It looked like I had slid over a food grater.

Even then, I would do anything to get at ripe berries. I imagine that, since the tree was in my aunt's backyard, I could have coaxed her to set up a stepladder. Then I could have picked a bucket of mulberries without resorting to climbing the tree. But I wanted those mulberries right now, not 5 or 10 minutes down the line.

My episode reminds me of the person praying to God. "Oh, Lord, grant me patience—right now!"

Mulberry, persimmon, papaw, and black cherry are the only trees listed in this book. The other wild berries and fruits are shrubs or vines seldom exceeding 20 feet in height.

You can pick mulberries by hand or you can spread tarps and shake the branches. The stem that adheres to the mulberry isn't of any concern—go ahead and cook it in your pies, jams, jelly, breads, and muffins.

One word of caution with mulberry: the raw shoots and green berries contain hallucinogens.

ॐ

IDENTIFICATION

Two species of mulberries occur in the Great Lakes area: red and white (*Morus rubra* and *Morus alba*). Red mulberry is native to the region. White mulberry was imported from Asia to foster silk growing; however, attempts were futile. White mulberry escaped cultivation and now grows wild throughout the United States.

The simple leaves are alternate. In both species, the sap is milky. Red mulberry leaves are 2 to 3 inches long, heart-shaped ovate, and broadest near the base. Sometimes one- to three-lobed, they are rough in texture above and downy underneath. White mulberry leaves are 1 to 3 inches long, coarsely toothed, and sometimes 2- to 3-lobed, with sparse hairs underneath the leaf.

Both species are dioecious—that is, a mulberry tree is either male or female. You will find mulberries only on the female trees. Mulberries blossom in late April and early May. Both male and female flowers appear on new growth. The female flower cluster resembles a small green mulberry in shape.

Mulberries ripen in late June into July. The berries look like blackberries in their size and cylindrical shape. The red mulberry turns from red to purple to nearly black. Ripe white mulberry is white to pinkish, but one variety produces dark red to black berries.

RECIPES

Spiced Mulberry Jam

I quart prepared mulberries
3 cups sugar
$1/4$ cup lemon juice
$1/2$ teaspoon cinnamon

Prepare mulberries by stemming and covering with cold salt water. Use $1/4$ cup salt to I quart water. Let stand 5 minutes. Drain. Rinse in cold water three times.

Crush berries. Add sugar, lemon juice, and cinnamon; cook slowly, stirring until sugar dissolves. Boil rapidly, stirring constantly to prevent scorching, until jellying point is reached.

Sheet test: Take a spoonful of hot jelly from the kettle and cool a minute. Holding the spoon at least a foot above the kettle, tip the spoon so the jelly runs back into the kettle. If the liquid runs together at the edge and "sheets" off the spoon, the jelly is ready.

Remove from heat; skim and stir alternately for 5 minutes. Ladle into sterilized jars; seal. Makes about three half-pints.

Betty Close, Kiel, Wisconsin

Rummy Mulberries

Wash and drain mulberries. Place in quart or pint jars. Cover with a rum-sugar solution. Seal tightly and store in a dark closet for three months before using. Especially good over ice cream. Rum-sugar solution: Dissolve 1 cup sugar in 2 cups potent rum.

Sharon Henry, Fort Smith, Montana

Mulberry Mousse

> 3 cups sugar
> 6 cups water
> 7$\frac{1}{2}$ cups crushed mulberries
> 6$\frac{1}{2}$ tablespoons lemon juice
> 3 cups heavy cream

Combine sugar and water and boil for 8 minutes. Force mulberry pulp and juice through a sieve. Add to sugar-water. Add lemon juice. Blend thoroughly. Place in a freezing tray and place in freezer. Stir 3 times at 30-minute intervals. Remove and fold in cream, beating 2 or 3 minutes if crystals are large. Return to tray and freezer for 2 to 3 hours longer.

Sharon Henry, Fort Smith, Montana

Mulberry Jam

Wash berries carefully, drain, and remove stems. For each pound of prepared fruit, allow an equal weight of sugar. Put berries in a 6- to 8-quart pot, crush, and bring slowly to a boil, stirring constantly. Add the sugar; boil until fruit mixture has thickened to jellylike consistency. Stir throughout the cooking. Pour into hot, sterilized jars and seal.

Sharon Henry, Fort Smith, Montana

Typical Blackberry
blossoms and fruit
(see page 15)

Typical Black Raspberry
blossoms and fruit
(see page 27)

Typical Black Cherry blossoms and fruit
(see page 22)

Typical Blueberry blossoms and fruit
(see page 34)

Typical Chokecherry blossoms and fruit
(see page 41)

Typical Cranberry
blossoms and fruit
(see page 47)

Typical Gooseberry
blossoms and fruit
(see page 62)

Typical Currant blossoms and fruit
(see page 54)

Typical Elderberry blossoms and fruit
(*see page 58*)

Typical Wild Grape
blossoms and fruit
(see page 68)

Typical Hawthorn
blossoms and fruit
(see page 72)

Typical Highbush
Cranberry blossoms
and fruit
(see *page 76*)

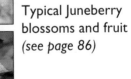

Typical Juneberry
blossoms and fruit
(see *page 86*)

STOWELL

Typical Huckleberry blossoms and fruit
(see page 80)

STOWELL & SKEAN

STOWELL & SKEAN

Typical Papaw blossoms and fruit
(see page 99)

STOWELL & SKEAN

STOWELL & SKEAN

Typical Mulberry
blossoms and fruit
(see page 93)

DOUG LADD

Typical Persimmon
blossoms and fruit
(see page 107)

CLARK SCHAACK

Typical Plum blossoms and fruit
(see page 112)

Typical Red Raspberry
blossoms and fruit
(see page 116)

Typical Wild Rose and
Rosehips
(see page 121)

Typical Strawberry
blossom and fruit
(see page 124)

Typical Thimbleberry
blossoms and fruit
(see page 129)

Red Baneberry
(poisonous, see page 11)

Woody Nightshade
(poisonous, see page 11)

Pokeweed
(poisonous, see page 12)

STOWELL & SKEAN

Common Buckthorn
(poisonous, see page 12)

STOWELL & SKEAN

Mulberry Slump

Batter:

3 tablespoons butter or margarine
4 tablespoons sugar
$1/2$ cup milk
$1 1/2$ cups flour
$1 1/2$ teaspoons baking powder
$1/4$ teaspoons salt

Cream first two ingredients together. Add milk and blend. Measure and mix dry ingredients in a separate bowl. Add dry ingredients to wet and mix. Drop batter by spoonfuls into bubbling berries. Cover and cook for 10 minutes.

Bubbling berries:

4 cups mulberries*
$1 1/2$ cups sugar
$1 1/2$ cups water
3 tablespoons cornstarch
1 tablespoon lemon juice

Bring to boil in heavy saucepan. Cinnamon or nutmeg may be added to taste. Simmer for 10 minutes before adding batter.

* Many berries can be substituted—for starters: blueberry, huckleberry, gooseberry, cranberry, and blackberry.

Charlotte Heron, Missoula, Montana

Mulberry Jam

8 cups crushed mulberries
6 cups sugar
$1/_4$ cup lemon juice
optional: powdered pectin

Combine in a large saucepan or kettle. Heat to boiling, stirring constantly. Boil until thick (about 10 to 15 minutes—you can take a shortcut by adding powdered pectin). Pour hot mixture into hot, sterilized jars. Seal. Process in hot water bath 10 minutes.

Charlotte Heron, Missoula, Montana

Papaw

Papaw, or pawpaw, is a member of a tropical group of trees, the custard apples, family Annonaceae. It grows in the southern portion of the Great Lakes, including the southern two thirds of Illinois, almost all of Indiana and Ohio, the southern third of lower Michigan and in far southern Ontario along Lake Erie, and through southern Pennsylvania and parts of New York and New Jersey.

The tree extends south in a broad band to northern Georgia and Louisiana, and west through Arkansas and southern Iowa.

Papaw has been called false banana, wild banana, or custard apple, because the stubby fruit looks somewhat like a small banana, and its flesh is rich, like custard.

The papaw is unique because it should be growing in the tropics, and because it doesn't go by the common rules of pollination. The blossoms of most plants use fragrances or color to attract such pollinating insects as bees, butterflies, and moths, but not papaw. Papaw's blossom attracts carrion insects, such as houseflies, beetles, and blowflies.

Dr. Stowell tells me of a fellow who grows papaws near Albion, Michigan. Every spring, the fellow gathers road kills and places them near his papaw grove so the carrion will attract insects and the blossoms will be fertilized! The fellow always gets papaws in the fall. I guess I shouldn't knock success, but at what price?

I find it amusing that an edible plant would have leaves that smell like motor oil when bruised. Perhaps here I should point out that not everyone likes papaw. It doesn't agree with some people's digestive systems.

Many people wait for the first frost before picking papaws so the flesh has a chance to soften. Pick papaws when they are yellow and just starting to soften. You can shake them from the branch, but remember, ripe papaws bruise easily, so handle them carefully. Store them in a cool place until they ripen fully. Then get ready for a tropical treat right in the Great Lakes area!

There seems to be quite a group of papaw cooks in Greenville, Ohio. According to Bob Welch, who sent me a number of recipes he and others like, "I have been told that papaws may be substituted in any recipe that calls for the use of bananas. One problem with using papaws in recipes is beating the raccoons to the ripe papaws!"

IDENTIFICATION

Papaw (*Asimina triloba*) is a small tree. It has dark brown bark with gray blotches and prominent whitish breathing pores (lenticels). The tree ranges from 10 to 40 feet in height and usually grows in groves (clones) in river valleys and rich bottomlands. Papaw wood is not strong: animals that climb papaws to obtain the fruit often break the tops of the trees out.

Papaw leaves are large—7 to 12 inches long—and simple, entire, and alternate. The dark green leaves turn yellow when the fruits are ripe.

The tree blossoms in late March and April. Blossoms occur along the stems of the preceding season's growth. At first, the bell-shaped flowers are a light green color, but they turn dark purple as they mature. The blossoms have two whorls of petals, the outer set considerably larger than the inner. Petals are conspicuously veined.

Technically, the papaw is a fleshy berry, cylindrically shaped, 2 to 5 inches long, and from 1 to 2$^1/_2$ inches in diameter. Papaw ripens in October, turning from green to yellow to deep brown or black when ripe. The pulp is orange and custardlike, with a pleasant, tropical fragrance. There are several large brown seeds in the pulp.

R E C I P E S

Papaw Pudding

> 3 beaten egg yolks (save the whites)
> I cup milk
> I cup cream
> $^1/_2$ cup sugar
> I cup mashed, strained papaw pulp

In a saucepan, combine beaten egg yolks, milk, cream, and sugar. Add strained papaw. Cook over low heat until hot. Refrigerate.

Before serving, top with meringue. Place in broiler just long enough to brown meringue; make sure the pudding remains cold.

Meringue:

> 3 egg whites
> $^1/_4$ cup sugar
> $^1/_2$ teaspoon cream of tartar

Beat egg whites with sugar and cream of tartar until stiff.

Sharon Henry, Fort Smith, Montana

Papaw Coffee Cake

$^1/_3$ cup shortening
$^2/_3$ cup sugar
2 eggs, well beaten
1$^3/_4$ cup flour
2 teaspoons baking powder
$^1/_4$ teaspoon baking soda
$^1/_2$ teaspoon salt
1 cup papaw pulp, seeds removed

Topping:

2 tablespoons sugar
$^1/_4$ teaspoon cinnamon
1 teaspoon freshly grated orange peel
$^1/_4$ cup chopped walnuts or pecans

Preheat oven to 350 degrees. Grease a 10-inch round cake pan. In a large bowl, cream shortening and sugar until fluffy. Add eggs and beat well. Sift flour, baking powder, soda, and salt together three times. Add dry ingredients to creamed mixture alternately with the papaw; start and end with dry ingredients. Mix well after each addition. Pour into prepared pan.

In a small bowl, prepare the topping by combining the ingredients. Sprinkle over the batter. Bake 35 minutes or until cake tests done.

Jean Buchner, Maquoketa, Iowa

Papaw Ice Cream

6 eggs
2 cups sugar
$^1/_2$ teaspoon lemon extract
$^1/_4$ cup vanilla
3 cups milk
3 12-ounce cans of evaporated milk
1 cup mashed papaw, seeds removed

In a large bowl, beat the eggs to a light yellow color. Gradually beat in sugar. Add lemon extract, vanilla, milk, and evaporated milk. Add papaw;

stir until thoroughly blended. Rinse container and dasher of a 1-gallon ice cream freezer and pour in mixture. Secure container in freezer tub and attach crank and gear assembly or electric motor head. Add 6 parts cracked ice to one part rock salt to within an inch of the lid. Add ice as necessary throughout freezing time, 20 to 25 minutes. Remove dasher. Drain off water. Repack with ice and salt. Let stand a couple of hours for ice cream to ripen. Yield: 1 gallon.

Toni Terrel, Wabash, Indiana

Papaw Pie

 1 cup sugar
 1 cup milk
 1 egg
 $^1/_4$ teaspoon salt
 $1^1/_2$ cups papaws peeled and seeded
 unbaked pie shell

Stir all ingredients together in a saucepan. Cook until thickened. Pour into pie shell and bake until done. Can top with meringue or other toppings.

Bob Welch, Greenville, Ohio

Papaw Pie

 6 to 7 ripe papaws, sliced, seeds removed
 8-inch double pie crust
 $^3/_4$ cup sugar
 $^1/_4$ cup apple cider
 3 tablespoons lime juice
 $1^1/_2$ teaspoons cinnamon
 $^1/_2$ teaspoons grated allspice
 $^1/_2$ cup heavy cream

Place papaws in 8-inch pie shell. Mix together sugar, cider, lime juice, spices, and cream; pour over papaws. Cover with top crust and bake for 50 minutes at 350 degrees.

Glenna Knoll, Greenville, Ohio

Baked Papaws

Select firm but ripe papaws. Bake in skins in a medium oven until done. Serve with sweet cream.

Glenna Knoll, Greenville, Ohio

Papaw Flump or Float

1 to 2 cups papaws, seeded
$1/2$ cup sugar
3 eggs whites
sweet cream
sugar

Mash papaws fine; add $1/2$ cup sugar. Beat egg whites until stiff and fold into papaw mixture. Bake 5 or 10 minutes in a medium oven. When cool, serve with sweet cream and sugar.

Glenna Knoll, Greenville, Ohio

Papaw Bread

$1/2$ cup shortening
2 cups flour
1 cup sugar
1 rounded teaspoon soda
1 teaspoon vanilla
1 teaspoon salt
2 eggs
$1/2$ cup chopped nut meats
1 cup papaw pulp, mashed, seeds removed

Cream sugar and shortening. Add vanilla, eggs, and pawpaw pulp. Stir well. Sift together flour, soda, and salt and stir into pawpaw mixture until the batter is smooth. Fold in chopped nuts. Bake for 1 hour in a 350-degree oven.

"This bread has a lovely pink color."

Bob Nixon, Greenville, Ohio

Papaw Preserves

12 papaws
1 orange
2 cups water
1 lemon
$^3/_4$ cup sugar

Peel papaws and cut into kettle without removing seeds. Add water and boil until soft. Put through a sieve; add sugar and juice from orange and lemon. Boil until thick, stirring constantly. The grated rind of the lemon or orange may be added if desired. Place in sterilized jars and seal. Process in hot water bath for 10 minutes.

Bob Nixon, Greenville, Ohio

Papaw Pudding or Chiffon Pie Filling

$^1/_2$ cup brown sugar
$^1/_2$ teaspoon salt
1 package Knox unflavored gelatin
$^2/_3$ cup milk
3 eggs
1 cup papaw pulp
$^1/_4$ cup of sugar
baked pie shell (if filling)

Mix brown sugar, salt, and gelatin. Add milk and 3 egg yolks. Cook mixture until it comes to a boil; stir in pawpaw pulp. Place mixture in refrigerator until chilled (20 to 30 minutes). Beat 3 egg whites, gradually adding $^1/_4$ cup sugar; mix until stiff peaks form. Fold egg whites into the pawpaw blend and serve as a pudding, or put in baked pie shell.

Susan Gray, Greenville, Ohio

Canned Papaws

Choose ripe but firm fruit. Wash and cut fruit and remove seeds. Make medium syrup (that is, bring 1 cup of sugar to 2 cups of water to a boil). After the syrup cools, add juice of 1 lemon or lime, and then sliced papaws. Bring slowly to a boil. Cook approximately 3 minutes. Pack in hot jars, adding enough hot syrup to cover. Adjust lids; process in a boiling water bath (for pint jars, 15 minutes).

Betty Rush, Greenville, Ohio

Persimmon

People associate persimmon with incredible sourness, but if you wait for persimmon to ripen, it is one of the tastiest of wild fruits.

Persimmon, a tree, is a member of the tropical family Ebenaceae, which includes many commercially important trees—for example, ebony. There are over 250 species in the family, and 7 genera. Only one genus, *Diospyros,* occurs in the United States, and only two species: common persimmon, *Diospyros virginiana;* and Texas persimmon, *Diospyros texana.*

Persimmon has an orange pulp and a flavor much like that of dates. According to my friends, Ted and Toni Terrel, it is best to wait until the leaves are off the trees before looking for persimmons. Most persimmons, Toni contends, are in the tops of the trees and the leaves obscure the view. Of course, that also means you might have to do some tree climbing to shake persistent persimmons out of a tree. Be careful. Ripe persimmons are tasty, but they aren't worth a fall and broken bones.

If you are persistent, the Terrels say, you can come up with a gallon or more of persimmons nine out of ten years. According to Ted, Thanksgiving is usually about the right time for ripe persimmons. But be quick: the raccoons and oppossums love them.

Toni mentions that you should not pick persimmons on horseback unless you have an appropriate container to put them in. One time she picked several, put them in her pocket, then trotted home. She ended up with a pocket full of persimmon pulp.

Persimmons can be used in breads, pies, and jams. You can freeze the pulp for later use. Make sure to strain it first to remove the skin and seeds.

IDENTIFICATION

In the Great Lakes region, common persimmon inhabits the Mississippi and Ohio River valleys—in other words, the southern thirds of Illinois, Indiana, and Ohio. The trees prefer the deep, rich soils of bottomlands. They grow up to 60 feet high and as wide as 2 feet in diameter. The thick, gray to grayish brown bark is rough and checkered. Branches are stout and spreading.

The somewhat leathery dark green leaves are alternate, simple, and entire. The leaves are rounded at the base, widest near the base or middle, and 2 to 5 inches long. They have abrupt, sharp tips.

Common persimmon is dioecious (meaning, roughly, "two houses"), so a tree is either male or female. What that means is that even in a good year, 50 percent of persimmon trees will not have fruits, because they are male. The tree's greenish yellow, urn-shaped flowers are situated in the angle of the leaf and branch.

Persimmon blossoms in the spring and ripens in the fall. The leathery calyx persists at the base of the persimmon. The persimmon, really a berry, is rather small compared to cultivated varieties: 1 to $2^1/_2$ inches in diameter. It is lustrous orange when ripe. After a frost or two, the smooth skin becomes wrinkled. That's the best time to pick persimmons! The four to eight reddish brown seeds are flattened and rounded at the ends.

R E C I P E S

Persimmon Bread

3 $^1/_2$ cups flour
1 teaspoon cinnamon
1 teaspoon nutmeg
1 $^1/_2$ teaspoons salt
2 teaspoons baking soda
1 cup cooking oil
2 cups persimmon pulp
$^2/_3$ cup water
4 eggs, lightly beaten
3 cups sugar
$^1/_2$ cup chopped nuts or raisins

Stir first 5 ingredients together; set aside. Mix oil, persimmon, and water with eggs; add sugar. Blend dry ingredients into egg mixture; add nuts. Divide batter into 4 greased 1-pound coffee cans. Bake at 350 degrees 1 hour. Yield: 32 servings.

Toni Terrel, Wabash, Indiana

Persimmon Pudding

1 cup persimmon pulp
1 egg, beaten
1 cup milk
1 teaspoon baking powder
$^1/_2$ cup sugar
$^1/_2$ teaspoon baking soda
2 tablespoons melted butter
1 $^1/_2$ cups flour

Measure persimmons after running them through a sieve. Stir in soda. Add sugar, beaten egg, and melted butter. Sift dry ingredients together and add alternately with milk to persimmon mixture. Bake 45 minutes at

(continued on next page)

300 degrees. The pudding will turn dark after it is removed from oven. Serve with whipped cream or pudding sauce.

Toni Terrel, Wabash, Indiana

Persimmon Cake

3 tablespoons shortening
3 teaspoons baking powder
$^3/_4$ cup sugar
$^1/_2$ teaspoon salt
1 egg, well beaten
$^1/_2$ cup milk
1$^3/_4$ cup sifted flour
$^3/_4$ cup persimmon pulp
confectioners sugar

Cream shortening, add sugar, then egg. Add dry ingredients alternately with milk; beat well. Add $^3/_4$ cup persimmon pulp. Grease a pan and line it with waxed paper; pour in mixture. Sift confectioners sugar over the top; bake at 350 degrees for 45 minutes to 1 hour.

Toni Terrel, Wabash, Indiana

Persimmon Pie

1 cup milk
1 tablespoon cornstarch
$^1/_2$ cup sugar
1 well-beaten egg
$^1/_4$ teaspoon salt
2 cups persimmon pulp
unbaked pastry shell

Combine ingredients in order given; pour into pastry shell. Bake at 450 degrees for 10 minutes, then reduce temperature to 350 degrees and bake 45 to 50 minutes longer.

Toni Terrel, Wabash, Indiana

Persimmon Pudding

1 1/2 cup sugar
2 1/2 teaspoons baking powder
1/2 cup butter
1 1/2 cup flour
3 eggs, well beaten
1 pinch salt
2 1/2 cups milk, cream, or half-and-half
1 pint persimmon pulp

Blend together sugar and butter. Mix in eggs, milk or cream, and persimmon pulp. Mix dry ingredients together, then add to first mixture. Place in an ungreased 9 x 13-inch pan (preferably Pyrex) and bake at 350 degrees for 40 to 50 minutes.

Lori Bishop, Wabash, Indiana

Persimmon Muffins

Combine in large bowl and let stand for 5 minutes:

2 cups 100 percent bran cereal
1 1/4 cup skim milk
1 cup wheat germ

Combine and sift into separate bowl:

1/3 cup brown sugar
2 teaspoons baking powder
1/2 teaspoon baking soda

Combine this third mixture:

1 egg
1 teaspoon vanilla
1 pint persimmon pulp

Mix all 3 combinations together. Fill lined muffin tins. Bake 18 to 20 minutes at 400 degrees. Makes 1 dozen muffins.

Lori Bishop, Wabash, Indiana

Plum

*W*ild *plum reminds me of a survivor in an adventure movie. Every year, wild plum withstands a variety of attackers: late frosts, insects, fungus, drought, and browsing deer. Yet the plant manages to survive the attacks, grows, and some years, produces a bountiful crop of tasty plums.*

While most wild berries supply a crop each year, and a bounteous crop every two or three years, wild plums bear a bumper crop about every three or four years. When you hit one of these wild plum bonanzas, you can pick enough plums to last a year or two.

Ranging from nickel- to quarter-sized, wild plums aren't anywhere near as large as domestic plums, but they can be just as tasty. The trick is to pick the plums when they are ripe, and to sample a lot of bushes. If you find a good thicket of wild plum and the plums are not rock-hard, you might want to pick them, then let them ripen fully in your basement. I have lost a bunch of plums waiting for them to be fully ripe; when I went out on the appointed day, I found that the raccoons or black bears had beaten me to them, or maybe a strong wind came up the day before and blew all the plums off the branches.

Dr. Stowell complained to me once that some wild plums he was trying to photograph were so ripe, when he would try to move a branch out of the way, the plums would fall!

Wild plums are a great snack food. I usually eat a few while I'm out fishing or bird hunting. They tend to grow in rocky or sandy soils along streams or woodlands, so many times, my fall travels bring me to plums. If plums aren't dead ripe, they won't be the best eating. Also, some patches of wild plums have superior fruit compared to other patches.

Your chances of getting a decent picking of wild plums are about one in four—but when you hit a good picking, it should be enough to hold you for a couple of years. Jam, jelly, liqueurs, butters, and fruit leather are just some of the yummy treats you can make with wild plum.

ᘓ

IDENTIFICATION

Wild plum prefers rich, well-drained, alluvial soils, but do well in gravelly soils, too. It grows in mixed deciduous forests and margins of fields, roadsides, and streams. In the Great Lakes area, you'll find wild plum, *Prunus americana*, in southern Minnesota, Wisconsin, and Michigan, and throughout Illinois, Indiana, and Ohio. Some wild plum grows in Ontario along Lake Erie.

Wild plum, a shrub or small tree, ranges in height from 3 to 25 feet. Its dark, grayish brown bark forms small plates, and its smaller branches have spines. Oftentimes you will find wild plum growing in thickets because it suckers freely—that is, new stems arise from the roots. The deciduous, alternate leaves are ovate to obovate with serrated margins.

Blossoms appear in May. About a $1/4$ inch across, they are white with five petals. Blossoms usually occur in clusters of three to five.

Wild plum ripens in September, turning from light green to deep pink or sometimes yellow. As mentioned before, the fruit is from nickel- to quarter-sized. The seed, a flattened pit, contains hydrocyanic acid, as do all members of the genus *Prunus*.

RECIPES

Wild Plum Jam

Wash and pit plums. Force through food chopper using coarse blade. Add 3½ cups sugar to 4 cups prepared plums; let stand 1 hour. Cook over moderately high heat, stirring frequently, until thick. Seal in hot, sterilized glasses.

Jeri Mazurek, Woodland, Michigan

Wild Plum Butter

3 quarts wild plums
½ cup water
9 cups sugar

Wash and sort plums, using only the ripest unblemished ones. In a 6- to 8-quart pot or saucepan place plums and water. Over low heat, bring to a slow simmer, until plums burst. Rub plum pulp through a fine sieve. Three quarts of plums should yield about nine cups of thick plum puree. Add sugar. Return to saucepan and simmer over low heat until thick. Pour hot plum butter into hot sterilized jars, seal and process in hot water bath for 10 minutes. Yield: about eight (8-ounce) jars.

Evelyn Hejde, Aladdin, Wyoming

Wild Plum Jelly

4 cups strained wild plum juice
3 cups sugar

Follow initial directions for making wild plum butter, but instead of passing the pulp through a sieve, use a jelly bag to extract the juice. Combine the juice with the sugar in a large saucepan or pot. Bring to a slow simmer over low heat until mixture reaches jell stage. (The sheet test is one method of determining if you have reached this stage—see "sheet test" in glossary.)

Evelyn Hejde, Aladdin, Wyoming

Wild Plum Preserves

Wash plums, place in kettle, cover with water, and cook until tender. Cool; with your hands, remove pits.* Mix fifty-fifty, by volume, with sugar (may use more or less sugar by personal taste). Cook till consistency is thick. Place in hot sterilized jars, seal and process in hot water bath for 10 minutes. Eat this covered with fresh cream from the separator—mmm good!

*A shortcut is to leave pits in and let the eater spit them out like watermelon seeds.

Mary Moravek, Sheridan, Wyoming

Plum-Raspberry Butter

1 1/2 pounds pitted plums
1 quart raspberries
1/2 cup water
1 1/2 cups sugar
1 tablespoon lemon juice

Place plums, raspberries, and water in a 6- to 8-quart pot or saucepan. Bring to a boil, reduce heat, and cook until tender, about 10 minutes. Cool. Puree in blender or food processor. Return to pot; add sugar and lemon juice. Cook over low heat until sugar is dissolved. Bring to boil; stir constantly until thick and glossy (about 10 minutes). Pour into hot, sterilized jars. Seal. Process in hot water bath for 15 minutes.

Charlotte Heron, Missoula, Montana

Red Raspberry

I magine hiking a wooded trail in the north country in midsummer. You still have an hour or two before you stop for lunch, but your stomach says you need a snack right now. Ahead of you is a clump of wild raspberries. Red, ripe berries beckon. Snack time is at hand.

Wild raspberry is one of the best snacks available to the outdoors person. A nice thing about these berries is that they usually grow at a comfortable height of 3 to 4 feet, so you don't have to bend over much to pick them— a handy thing with a full backpack on your shoulders!

So many times, wild raspberries have given me a snack break, whether I was on a weeklong pack trip or a day hike, or fishing for trout in northern Michigan. While it might be a two-hour job to pick enough small raspberries to cook with, five minutes of picking can give me more than enough food to last until mealtime.

Wild raspberry is about twice as luscious in taste as tame varieties. This taste inequality is true for just about any wild variety of berry or fruit versus cultivated. Perhaps nature contrived to add more flavor so animals would seek out the berries and fruits and distribute the seeds.

Though wild raspberry usually produces well, getting a picking of two quarts or more is chancy, because other critters like the berries too; finding a patch that's big enough can be hard to impossible. Still, a handful of fresh wild raspberries is a heck of a lot better than a sharp stick in the eye.

ౠ

IDENTIFICATION

Wild raspberry likes recently disturbed sites and the edges of woodlands and streams. Recent timber cuts are a good place for raspberry to take up residence. The best concentrations of wild raspberry I have found have been along abandoned roads leading into strip cuts.

Rubus idaeus is the most common wild raspberry in the Great Lakes area. It is found in all of Michigan, in eastern Indiana, and in most all of Ohio, Ontario, and Quebec.

Raspberries and blackberries have a growth form called a cane. A cane has a woody exterior, but a pithy interior. While the roots are perennial, canes live but two years (that makes them biennial). Their first year, canes are unbranched, and have prickles and compound leaves. Second-year canes have small branches which blossom and bear fruit.

Wild raspberry canes are light brown (although first-year canes, still growing, are very light green in color), with numerous to slight numbers of prickles. Canes are from 2 to 5 feet high. Leaves are compound with 3 to 5 leaflets. On the surface they are dark green to light green; underneath, they are light green to whitish, with pubescence.

Wild raspberry blossoms from May to July, depending on the latitude. The flower has a 5-part calyx and 5 white petals. Flowers are about $3/8$ of an inch in diameter and in clusters.

Wild raspberry ripens from July to September, depending on the latitude. Surrounded by a light green, 5-part calyx "collar," the berry turns from light green to red. When you pick a red or black raspberry, the receptacle, a white, cone-shaped part, remains on the bush. When you pick blackberries or dewberries, the receptacle comes with the berry.

Technically, the raspberry fruit is not a berry, but an aggregation of small drupes. Tough coats on the seeds mean that all members of the genus *Rubus* have chewy berries.

RECIPES

Raspberry Cake

2 cups flour
1 egg
$^1/_2$ teaspoon salt
1 teaspoon vanilla
3 teaspoon baking powder
1 cup milk
$^1/_3$ cup butter or margarine
2 cups raspberries*
1 cup sugar

Mix flour, salt, and baking powder. Cream butter and sugar; add egg. Add dry ingredients alternately with milk and vanilla to butter mixture. Pour into greased 9 x 13-inch pan; sprinkle with raspberries. Bake at 375 degrees for about 30 minutes. While cake is still warm, frost with the following:

1$^1/_2$ cups powdered sugar
1 teaspoon melted butter
3 tablespoons cream or milk

*Black raspberries or blackberries can be substituted.

Betty Close, Kiel, Wisconsin

Red Raspberry Kuchen

Crust:

1 cup flour
3 tablespoons powdered sugar
$^1/_2$ cup butter
1 to 1$^1/_2$ cups raspberries

Mix together and pat into bottom of a greased 9 x 13-inch pan. Add raspberries in a single layer. Pour the following mixture over the raspberries:

1 cup cream
1 cup sugar
1 heaping tablespoon flour

Bake at 350 degrees about 1 hour.

Betty Close, Kiel, Wisconsin

Grandma Lydia Kemper's Berry* Good Recipe

1$^1/_2$ cups raspberries (approximately)
1 cup flour
1 teaspoon baking powder (heaping)
$^1/_4$ to $^1/_2$ cup sugar
pinch salt
$^1/_2$ to 1 teaspoon vanilla
1 small egg
$^1/_4$ cup thick cream or evaporated milk
enough milk to make dough

Put raspberries in a thick layer in a 9 x 13-inch greased baking dish. Sprinkle an appropriate amount of sugar on top. Mix rest of ingredients together; spread over raspberries. Bake at 350 degrees until nicely brown.

*You can substitute black raspberries, blackberries, blueberries, mulberries, or juneberries.

Grandma Lydia Kemper, Crete, Nebraska

Raspberry-Cranberry Ring

I 3-oz package raspberry gelatin
I 3-oz package lemon gelatin
2 cups boiling water
I quart red raspberries
I cup cranberry-orange relish*
I cup lemon-lime beverage

Dissolve raspberry and lemon gelatin in boiling water. Cool; add raspberries and cranberry-orange relish. Chill until cold but not set. Carefully, pour in lemon-lime beverage. Stir gently. Chill until partially set. Turn into a 5 1/2-cup mold. Chill until firm—about 4 hours. Unmold on crisp greens.

*See recipes in cranberry section of this book.

Jeri Mazurek, Woodland, Michigan

Rosehip

Wild rose has some great pluses: it has such fragrant, delicate blossoms, and its fruits, called "hips," have the highest natural concentration of vitamin C of any plant — on a weight basis, about one hundred times more than that of an orange. In addition, rosehips make tasty tea and delicate jelly.

That's not bad for a shrub that will leave you bleeding if you try to walk through it. Wild rose provides cover and food for game birds like pheasant, sharp-tailed grouse, and turkey, in addition to giving songbirds nesting sites.

I've had some tough times with wild rose when I'm out bird hunting, but I usually don't bear the plant any malice. After all, the same thorns and prickles that deter me put the skids on foxes and coyotes, offering sanctuary for the birds. Besides, rosehip rinds give me something to snack on when I'm hunting or hiking, though sometimes I have to try several bushes before I find one with an acceptable taste.

Ripening in September, wild rosehips can be used throughout winter as survival food. Most years, crops are plentiful. Four out of five years, you should be able to pick a gallon or two. Collect hips in the fall, preferably after the first frost, when they are still firm, but red and ripe. To retain their vitamin C content, prepare rosehips soon after collecting.

IDENTIFICATION

Wild rose, represented by a large number of species all belonging to the genus *Rosa*, grows throughout the Great Lakes area. Wild rose likes sunny, open areas, so you can find it along roadsides, field edges, the perimeter of woods, and recently disturbed sites. Wild rose blossoms in June or early July. Its large blossoms, the size of a quarter to a silver dollar, have five petals usually pink in color, but sometimes rose-purple or white. Rose blossoms have numerous stamens and pistils.

Almost all rose bushes are 2 to 6 feet tall, though some species seldom exceed 1 foot, and others grow to 10 feet in height. Prickles or small thorns are characteristic of rose, though some species have very few prickles, and others could stop a charging buffalo with their armament.

Leaves are arranged alternately and are pinnately compound. The number of leaflets varies from three to as many as nine, though with most species, the number is five to seven. Leaflets are broadest near the base and have pronounced teeth.

The hips are orange- or red-colored with a five-pointed calyx opposite the stem. The sepals that make up the calyx vary in length from a 1/2 inch to an inch long.

RECIPES

Rosehip Jelly

Wash and remove the "tails"; cover with water, bring quickly to a boil, and cook slowly for 15 minutes. The juice, extracted, bottled, and pasteurized, can be stored in a dark cool place to be used later. The pulp, sieved to remove seeds and skins, can be used in jam, marmalades, and ketchups.

4 cups rosehip juice
7¹/₂ cups sugar
I packet of liquid pectin

Measure juice and stir in sugar. Place on high heat and, stirring constantly, bring to a quick rolling boil that cannot be stirred down. Add pectin and bring to a full rolling boil. Boil hard I minute. Remove from heat; skim off foam. Pour jelly immediately into hot, sterilized containers and seal. Process in hot water bath for 10 minutes.

Pauline Deem, Plentywood, Montana

Dried Rosehips

Cut rosehips in two and remove seeds with the point of a knife. Dry as quickly as possible in a slightly warm oven or food dehydrator. Crumble dried hips and use as desired.

"Dried rosehips can be used to sprinkle over desserts or cereals, added to batters for baked goods, or combined with tea or fruit juices or a cold beverage."

Pauline Deem, Plentywood, Montana

Rosehip Syrup and Fruit Butter

Snip bud ends from hips. Cover fruit with water and boil until soft. Cool. Grind in blender. Strain juice. For every 2 cups juice, add I cup sugar. Boil until thick. Bottle or seal in jars.

Press pulp through a sieve. For 2 cups pulp, add I cup sugar. Add cinnamon, cloves, and allspice to taste. Heat to dissolve sugar. Uncover and cook, stirring constantly to prevent sticking. Pack in jars and seal. Process in hot water bath 10 minutes.

Charlotte Heron, Missoula, Montana

Strawberry

Wild strawberry (Frageria sp.) occurs throughout most of eastern North America. That fortuitous circumstance means that you can enjoy some great trail snacks anywhere from Hudson Bay to southern Alabama.

One of my fondest memories of wild strawberries involves a June trout fishing trip my dad took with my brother and me to north-central Michigan. We were camped on a tributary of the Little Manistee River. My brother and I enjoyed some good fishing early in the morning, but by midday, as the heat increased, the fishing turned off.

With nothing else to do, I elected to explore a pine barrens area. I had a bit of good luck: I stumbled onto a rather large strawberry patch on the edge of the barrens. After eating several handfuls, I thought I would pick some for camp breakfast, so I hiked back to get a pan to put the berries in. At camp, Dad was lounging in the shade. I told him I had found some wild strawberries and was going to pick some for breakfast. He told me, "Don't bother, they are too small, you'll not pick enough to make it worthwhile."

I was a teenager and didn't take advice like that very well. I went ahead anyway. The upshot of the story is that I picked 2 or 3 cups of luscious berries. The berries were plenty enough to cover our cereal, plus give us a small bowl each of berries and milk.

Dad had to admit, the strawberries were a great treat, and well worth my effort. That patch provided us with yet another breakfast during our fishing trip.

While wild strawberries may not yield a lot of volume, they sure have high-volume taste. Today's domestic strawberry varieties offer huge berries, but oftentimes the taste is not commensurate with the size.

Wild strawberry is fairly well distributed throughout the Great Lakes area, but seldom do you find a patch big enough to yield more than two or three cups of berries. I would rate your chances of finding enough to make jam (2 quarts) about one in ten or less.

IDENTIFICATION

Wild strawberry blossoms in May and bears fruit in June or early July. In southern areas of the Great Lakes states, look for the bright white blossoms with equally bright yellow centers in early May. Ripe berries appear in early June. Ripe wild strawberries look like their cultivated cousins, only smaller. A big one is the size of a penny; normal size is slightly smaller than a dime.

Wild strawberry doesn't have woody stems; rather, this herbaceous perennial produces new stems and leaves each year. The plants are 3 to 4 inches high. Most strawberry plants produce runners. All the leaves originate from a basal clump. The light green leaves are composed of three leaflets with prominent teeth. Underneath, the leaflets have small hairs.

RECIPES

Black Cherry-Strawberry Jam

2$\frac{1}{2}$ cups pitted black cherries
$\frac{3}{4}$ cup water
2 cups fresh strawberries
1 package pectin
4$\frac{1}{2}$ cups sugar

Grind or finely chop cherries; measure 1$\frac{1}{2}$ cups. Wash and stem strawberries; mash; measure 1 cup. In large bowl combine cherries, strawberries, and sugar. Let stand 10 minutes. In small saucepan, combine water and fruit pectin. Bring mixture to a full rolling boil. Boil hard 1 minute, stirring constantly. Stir hot pectin into the fruit mixture; stir for 3 minutes. Ladle at once into clean jars or jelly glasses, or freezer containers; seal and label. Let stand for several hours at room temperature or till jam is set. Store jam in the refrigerator for 3 weeks or in the freezer up to 3 months.

Emily Krumm, Eaton Rapids, Michigan

Strawberry Ice Cream Pie

1 baked 9-inch graham cracker crumb crust
1 quart strawberry ice cream
1 pint fresh strawberries
whipped heavy cream

Let ice cream soften slightly at room temperature. Spread into cool crumb crust. Place in freezer. To serve, cut in wedges. Garnish with whipped cream and ripe strawberries, sweetened.

Jeri Mazurek, Woodland, Michigan

Thick Strawberry Shake

1 cup milk
1 cup ice cream or ice milk
1 cup wild strawberries*
$1/_4$ cup sugar (you may need a little more)

Place all ingredients in an electric blender. Blend until smooth. You may need a spoon to eat this shake with; if you want it a little thinner, use more milk.

*Other wild berries may be used: red and black raspberry, blackberry, blueberry, thimbleberry, and mulberry.

Sharon Henry, Fort Smith, Montana

(more strawberry recipes on next page)

Strawberry Creme Pie

Pie shell:

1 cup flour
$^1/_2$ teaspoon salt
$^1/_2$ cup shortening
enough water to make dough

Bake pie shell at 425 degrees for approximately 10 minutes or until done.

Filling:

1 cup sugar
6 tablespoons cornstarch
$^1/_2$ teaspoon salt
$2^1/_2$ cups milk
2 beaten eggs
$^1/_2$ teaspoon vanilla
1 pint strawberries (at least)

Mix sugar, cornstarch, salt, milk. Cook over medium heat, stirring constantly so it doesn't scorch (you can use a double boiler). When it thickens, add a small amount of the mix to the beaten eggs; stir that into the mix. Remove from heat. Add vanilla. Cool.

Line baked pie shell with strawberries. Pour filling over berries. Add more berries on top (in a decorative pattern if you wish). Chill.

Charlotte Heron, Missoula, Montana

Thimbleberry

T himbleberry strikes me as a berry that couldn't figure out what it really wanted to be when it grew up. Thimbleberry, Rubus parviflorus, is in the same genus as raspberries and blackberries, but the plant looks like a maple bush. The large maplelike leaves are the size of saucers. The blossom is big enough to be a daisy.

Still, when a thimbleberry ripens, there is no doubt it is kin to raspberry. The berry looks like a big, deep red raspberry. When you taste a thimbleberry for the first time, you will think some plant scientist stumbled across a formula for intensifying raspberry flavor tenfold.

Unfortunately, thimbleberry is not widespread in the Great Lakes region. It only occurs in Michigan's upper peninsula, the northern fourth of the lower peninsula, far northern Minnesota, far southwestern Ontario, and southern Manitoba. However, thimbleberry is widespread in the Rocky Mountain West and the Northwest.

Thimbleberry ranks high on the flavorful list; if you like berries, make it a point to pick some of these. Thimbleberry goes well as is; with milk and sugar; and in pancakes, jam, and jelly. Basically, any way you would eat red raspberry is a good way to eat thimbleberries.

Your chances of getting 2 quarts of thimbleberry are dicey, because there are so few flowers per cane; but I have seen patches that run for 4 or 5 acres. If you find a big patch like that, your chances are good. If you are new to thimbleberry picking, I would rate your chances as one in ten; experienced thimbleberry pickers rate a one-in-three chance.

IDENTIFICATION

Thimbleberry likes moist, well-drained, rich soils in open forests or along trails and streams. There are no prickles on thimbleberry canes; the growing part of the canes are light green in color, while older parts have gray shredded bark. Thimbleberry attains heights of 2 to 6 feet. Its large, alternate, three- to five-lobed leaves are maple-shaped. The light green leaves, coarsely toothed, are pubescent on both surfaces.

The flowers, 1 to 2 inches wide, have five white outer petals and a yellow center, much like a big strawberry blossom. Five light-green sepals are prominent, and longer than the petals. There are few (3 to 6) flowers in each cluster. Thimbleberry blossoms in mid-June to mid-July.

The scarlet red hemispheric berries ripen from August into September. Just as with raspberries, the receptacle stays on the plant when you pick it.

R E C I P E S

Thimbleberry Shrub

I cup fresh or frozen thimbleberries*
$^1/_2$ cup ice water
I slice lemon, with rind, $^1/_4$ inch thick
$1^1/_2$ teaspoons honey
sprig of mint

Combine thimbleberries, water, lemon rind, and honey, and process in a food blender until smooth. Chill. When ready to serve, pour into a sherbet or parfait glass. Garnish with mint sprig.

* It is unnecessary to thaw the thimbleberries, since a shrub should be served very cold. Red or black raspberries can be used as well.

Kathy Krumm, Jackson, Michigan

Thimbleberry Jam

Wash and pick over thimbleberries. Measure. Combine with two thirds of that amount of sugar. Stir mixture over moderate heat until sugar dissolves. Turning to high heat, bring mixture to a boil until it has a thick consistency and the berries become clear. Stir constantly. When the jam sheets off a spoon, the jell stage has been reached. Pour into hot, sterilized jars; put on lids and bands. Invert for 5 minutes. (Or use USDA hot water bath for 10 minutes.)

**Old recipe, author unknown. Submitted by Mary Minnich,
Eaton Rapids, Michigan**

Thimbleberry-Currant Sherbet

2 pints fresh thimbleberries
1 1/4 cups currant jelly
2 cups half-and-half or light cream
1/2 cup creme de cassis

Puree thimbleberries in blender. Place the puree and jelly in a 2-quart pan. Heat on low, stirring often, until jelly melts. Cool to lukewarm. Stir in half and half and creme de cassis. Chill 1 hour. Churn-freeze. Place in plastic freezer container. Let ripen 3 hours in freezer before serving.

Charlotte Heron, Missoula, Montana

Thimbleberry Jam

6 cups crushed thimbleberries
1 package powdered pectin
8 1/2 cups sugar

Crush approximately 3 quarts of thimbleberries in a 6- or 8-quart saucepan or pot. Measure 6 cups. Stir pectin in thoroughly. Measure sugar into a bowl and set aside. Bring crushed berries to a rolling boil, stirring constantly. Stir in sugar; stirring constantly, bring to a full rolling boil. (You may add 1/4 teaspoon margarine or butter to minimize foaming at this point.) Boil for 4 minutes. Skim off foam immediately and pour into hot, sterilized jars. Wipe off rims, seal, and process in a hot water bath for 10 minutes.

Charlotte Heron, Missoula, Montana

Berry Good Hints

Berry Good Hints: Cross Reference of Berries' Best Uses

SNACK BERRIES

These are berries you can eat "as is," without any preparation.

Blackberry
Blueberry
Dewberry
Huckleberry
Juneberry
Mulberry, white and red
Raspberry, black and red
Wild Strawberry
Thimbleberry

PIE BERRIES

Blackberry
Black cherry
Blueberry
Chokecherry
Currant
Elderberry
Gooseberry
Huckleberry
Juneberry
Mulberry
Raspberry, black and red
Wild Plum
Wild Strawberry

VERSATILE RECIPES

You can substitute one or more kinds of berries in these recipes.

Jelly Roll (Blackberry chapter)
Old-Fashioned Berry Fritters (Black Raspberry chapter)
Hobo Cookies (Black Raspberry chapter)
Blueberry Sourdough Pancakes (Blueberry chapter)
Blueberry Pancake Syrup (Blueberry chapter)
Gooseberry Slump (Gooseberry chapter)
Highbush Cranberry Milk Shake (Highbush Cranberry chapter)
Huckleberry Pancakes (Huckleberry chapter)
Juneberry Muffins (Juneberry chapter)
Mulberry Slump (Mulberry chapter)
Grandma Lydia Kemper's
Berry Good Recipe (Red Raspberry chapter)
Thick Strawberry Shake (Strawberry chapter)
Thimbleberry Shrub (Thimbleberry chapter)

CAMP RECIPES

These are recipes you can make over a camp stove or over a campfire.

Gooseberry Slump (Gooseberry chapter)
Blueberry Sourdough Pancakes (Blueberry chapter)
Huckleberry Pancakes (Huckleberry chapter)
Blueberry Pancake Syrup (Blueberry chapter)
Mulberry Slump (Mulberry chapter)

Calendar

	Blossoms	Ripens
Blackberry	mid-May to mid-June	mid-July–August
Black Cherry	May	August–September
Black Raspberry	May–June	July–August
Blueberry	June	August–September
Chokecherry	May	August–September
Cranberry	June–July	September–October
Gooseberry	late April to mid-May	late July–August
Hawthorn	mid-May to mid-June	late August–September
Highbush Cranberry	mid-June	October–winter
Huckleberry	late May–June	August–mid September
Juneberry	late April–mid-May	mid-July–mid-August
Mulberry	late April–mid-May	early June–July
Papaw	April into May	mid-September–October
Persimmon	April into May	October–November
Plum	late April–May	September
Red Raspberry	June	late July–August
Strawberry	mid-May–June	June–July
Thimbleberry	May–June	June–July
Wild Grape	mid-to late May	September
Wild Rose	June	September

A plant will blossom earlier at lower latitudes and later at higher latitudes. The same holds true for when the berries or fruits ripen.

Glossary

Alternate: a type of leaf arrangement with only one leaf per node on alternating sides of the stem.

Biennial: describes a life cycle that completes in two years.

Blush: a whitish, powdery covering of the fruit, berry, leaf, or twig.

Bract: a small leaf beneath a flower, or belonging to a flower cluster.

Calyx: the green outer whorl of a flower composed of sepals

Cane: a pithy stem such as is found in raspberry or elderberry.

Clone: a grouping of plants which have the same genetic material due to asexual means of reproducing. Raspberry, blackberry, and blueberry plants often replicate from root stock, thus the patch formed from the same root stock is a clone.

Corymb: a round-topped or flat-topped blossom in which the outer flowers blossom first.

Cultivar: a cultivated variety of plant.

Cyme: a flat-topped or nearly flat-topped flower cluster.

Dioecious: having the female flowers on one plant and the male flowers on another, e.g., persimmon.

Drupe: a fleshy fruit with a pit or stone, e.g., cherry or plum.

Entire: without divisions, lobes, or teeth.

Escape: a cultivated variety of plant that has escaped cultivation and occurs in the wild, e.g., wild asparagus.

Glabrous: completely smooth, without hairs or bristles.

Hemispheric: roughly half of a sphere or round object.

Hip: a fleshy, cuplike receptacle that is the fruit of the rose.

Inferior Ovary: the flower parts arise from the top of the ovary.

Lanceolate: lance-like or shaped like a spear point. A long, narrow leaf that is several times longer than it is wide.

Lenticels: breathing pores in the bark that resemble warts or light-colored spots.

Mesic: the midrange of an environmental variable, e.g., moisture.

Obovate: inversely ovate.

Opposite: a type of arrangement in which two leaves grow on each side of a stem at each node. The leaves are opposite one another. Contrasts with alternate leaf arrangement.

Ovate: having the shape of a longitudinal section of a hen's egg in outline.

Palmately Compound Leaf: the leaflets arise from a central point.

Peduncle: the part of a stem which bears an inflorescence or single flower, either leafless or with bracts.

Pinnately Compound Leaf: the leaflets arise along a central stem.

Pome: fleshy fruit from an inferior ovary, e.g., apple.

Pubescent: covered with soft hairs, downy.

Raceme: an inflorescence or cluster of flowers along one main stem.

Recumbent: lying on or leaning close to the ground. Dewberries are often recumbent.

Render: to extract juice or pulp (cooking term).

Rolling Boil: a boil that does not stop when stirred (cooking term).

Runner: a horizontal, above-ground stem which roots at the nodes, producing a new plant.

Sepal: a part of a flower situated beneath the petals. The sepals are often green-colored.

Serrate: having small teeth.

Sheet Test: a method of determining the point at which jelling occurs. Take a spoonful of hot jelly from the kettle and cool a minute. Holding the spoon at least a foot above the mixture, tip the spoon so the jelly runs back into the kettle. If the liquid runs together at the edge and "sheets" off the spoon, the jelly is ready (cooking term).

Simple: describes a leaf with a single blade.

Stamen: the male reproductive part of the plant that produces pollen.

Stigma: the female part of a flower that receives the pollen.

Tendril: an outgrowth at the end of a leaf or stem node used for clinging or climbing.

Venation: the system of veins in the tissue of a leaf.

Whorl: a circular arrangement of petals or leaves around a stem.

Bibliography

Arnold, Robert E. *Poisonous Plants*. Jeffersontown, Ky: Terra Publishing, Inc. 1978.

Barnes, Burton V. and Warren H. Wagner, Jr. *Michigan Trees*. Ann Arbor: University of Michigan Press. 1981.

Billington, Cecil. *Shrubs of Michigan*. Bloomfield Hills, Michigan: Cranbrook Institute of Science Bulletin No. 2, 2nd edition, 1949.

Elias, Thomas S. *The Complete Trees of North America*. New York: Book Division, Times Mirror Magazines, Inc., 1981.

Elias, Thomas S., and Peter A. Dykeman. *Field Guide to North American Edible Wild Plants*. New York: Outdoor Life Books, 1982.

Gleason, Henry A., and Arthur Cronquist. *Manual of Vascular Plants of the Northeastern United States and Adjacent Canada*. 2nd edition. Bronx: New York Botanical Garden, 1991.

Hall, Alan. *Wild Food Trail Guide*. New York: Holt, Tinehart and Winston, 1973.

Lewistown Chamber of Commerce. *Montana Chokecherry Festival Recipe Book*. Lewiston, Montana: Lewistown Chamber of Commerce, 1991

Stokes, Donald W. *Natural History of Wild Shrubs and Vines: Eastern and Central North America*. New York: Harper and Row, 1981.

Sutton, Ann, and Myron Sutton. *Eastern Forests*. New York: Knopf, 1985.

Voss, Edward G. "Michigan Flora, Part II, Dicots. Cranbrook Institute Science Bulletin 59 and University of Michigan Herb." 1985

Weatherbee, Ellen Elliott, and James Garnett Bruce. *Edible Wild Plants of the Great Lakes Region*. University of Michigan: published by the authors, 1979.

Woodward, Lucia. *Poisonous Plants: A Color Field Guide*. New York: Hippocrene Books, 1985.

Young, Darrell D. *Wild Plants You Can Eat*. New York: Julian Messner, 1983.

Common & Scientific Names Index

S

Sambucus canadensis 58–61
Sambucus pubens 59
Saskatoon berry. See Juneberry
Saskatoon serviceberry 88
Serviceberry. See Juneberry
Shadbush. See Juneberry
Solanum dulcamara 12
Strawberry 54, 87, 124–128
Swamp sugar pear. See Juneberry

T

Thimbleberry 18, 29, 129–132

V

Vaccinium 76, 80
Vaccinium angustifolium 35
Vaccinium corymbosum 36
Vaccinium macrocarpon 47–53
Vaccinium myrtilloides 36
Vaccinium vacillans 36
Viburnum 76
Viburnum opulus 76
Viburnum trilobum 76–79
Vitis riparia 69

Recipe Index

Alaska's Bountiful Berries

Grab your pail and head for the hills.
It's berry-picking time!

BY JANICE J. SCHOFIELD

This time of year, the northland delivers up a bounty of plump, juicy gems. Bruin, bird and human alike race to the fragrant berry patches to share in the harvest.

Alaska is a berry-picker's nirvana. This wide-open land proffers an unmatched variety of edible berries, from the bog blueberry, which grows nearly everywhere, to the salmonberry, which is a little more selective about where it takes root.

So many berries, so little time. Anticipation drives human berry pickers rabid. With whatever container is available—from bucket to baggie—they race to the fields and hills to scoop up the waiting fruit.

After a few frenzied hours, the inevitable high—contagious and irreversible—sets in. No amount of back pain or finger tingling can stop the serious foragers now. "Just one more berry," they mumble, eyes probing the tangled foliage ahead. The mother lode is always one bush, one crest away.

Success is sweet. Flushed, and with juice-stained hands and knees, the afflicted finally realize they can't pick every berry on the hill and proudly carry their spoils home. They'll enjoy the harvest, topped with cream perhaps, or mounded on top of cereal, folded into cobbler, dropped into pancakes.

The following feature identifies eight of the more popular edible berries found in Alaska and was adapted from author Janice J. Schofield's book, *Discovering Wild Plants.*

Read it, and then grab your pail. It's berry-picking season, so let the gathering begin.

▼ Raspberry
Rubus idaeus

Raspberries are found in fields and thickets from the Interior to Southcentral and southeastern Alaska. Indescribably delicious, raspberries also are laden with vitamins B and C and the minerals magnesium, calcium, iron and phosphorus.

According to Chinese herbals, eating raspberries clears the complexion, cures colds, increases virility (or fertility), stimulates hair growth and heightens the body's vigor. When savoring raspberries, you can rest assured you are "treating" yourself in more ways than one.

Harvest the red fruit during the summer, generally in July or August, depending on local conditions.

◄ Lowbush Cranberry (Lingonberry)
Vaccinium vitis-idaea

The lowbush cranberry thrives in acid soils throughout Alaska. Its habitat varies from bogs and open woods to tundra.

Harvest the red berries during the fall. They're best after the first frost. Store them using the centuries-old system of placing them in a crock, covered with water, in a cool place. Or place firm, unbruised berries in containers in your freezer.

▼ Currant (Gooseberry)
Ribes species

Currants thrive in moist woods, open areas, stream banks, meadows, roadsides and logged areas. Many species exist in Alaska, most notably the northern red currant, *R. triste*, which has an exceptional flavor without the skunklike smell associated with many currants.

Currants are harvested in mid- to late summer. Fruits are best for preserves when slightly underripe.

RAY HAFEN/KEN GRAHAM AGENCY

VERNA PRATT

VERNA PRATT

◄ Blueberry
Vaccinium species

Blueberry habitat varies with species, ranging from thickets, woods and moist meadows to bogs and tundra. A number of blue-fruited *Vaccinium* species are referred to as blueberries, including the oval-leaf or early blueberry (*V. ovalifolium*), bog blueberry (*V. uliginosum*), dwarf blueberry (*V. caespitosum*, also known as *V. nivictum*), and Alaska blueberry (*V. alaskensis*).

Although the harvest time varies with species and local conditions, blueberry-picking season generally is in early to late summer.

FRED HIRSCHMANN

Author JANICE J. SCHOFIELD, a former home economics teacher, is an authority on Northern wild plants and their many uses. She lives in Homer, where she owns Gardensong Herbs and instructs herbal classes.

NOTE: *This feature looks at some of the most popular edible berries found in Alaska. Although we made every effort to include clear, identifying photographs and information, this excerpt is not intended to replace a guidebook. There are many poisonous plants in Alaska, including some that produce berrylike fruit. Please take the time to carefully identify any berries before eating them. When in doubt, consult an experienced local forager or the experts at the Alaska Cooperative Extension in Anchorage at (907) 279-5582.*

FRED HIRSCHMANN

◄ Salmonberry
Rubus spectabilis

Thickets of salmonberries are common in moist woods and lower mountainous regions, and along roadsides, ranging from the Alaska Peninsula and Southeast to Southcentral.

Cousin to the raspberry, the fruits look similar, although salmonberries may be either red or orange at maturity. In the north, Alaskans often refer to cloudberries, *R. chamaemorus*, as salmonberries, which don't grow in the Arctic. While both are members of the *Rubus* genus, the cloudberry's hard red fruit is smaller, and turns plump and golden at maturity. Also, the salmonberry is a stately shrub, while the cloudberry grows only a few inches high.

However, the confusion is understandable. Immature salmonberries pose a striking resemblance to the mature cloudberry.

Harvest the berries during the summer (the fruit ripens early in coastal areas, later in the mountains).

▼ Highbush Cranberry
Viburnum edule

Woods, riverbanks and thickets are the highbush cranberry's habitat throughout much of Alaska. Despite its name and red fruit, *V. edule* is unrelated to the Thanksgiving condiment. The true cranberry belongs to the heath family, whereas the highbush cranberry is a member of the honeysuckle family.

Harvest the berries in late summer or early fall. (The proper time for harvesting is a matter of controversy among foragers. The fruits, which are hard and exceedingly sour before frost, become soft and mildly acidic after exposure to chilly autumn evenings. As they age and sweeten, however, they develop a musty odor some find objectionable. Though definitely better as a trail snack after a frost, highbush cranberries often are picked before, when slightly underripe, for the making of what some consider superior preserves.)

LOREN TAFT

ERNEST MANEWAL

▲ Prickly Rose rose hips
Rosa acicularis

Found throughout much of Alaska, the prickly rose thrives in habitats as variable as meadows, forests, mountain slopes and bogs. The rose hip, or fruit, is harvested after the first frost. Rose hips are renowned for their nutritional content. Just three marble-sized hips have more vitamin C than a whole orange. In addition, they're high in vitamins A, B, E and K.

To seed or not to seed rose hips depends greatly on your personal preference and intended use. Generally, you'll want to at least remove the tails (withered sepals) from the hips. You then have the option of using the hips whole or halved, with or without seeds. To deseed hips, slice them in half and scoop out the seedy-hairy interior.

◄ Crowberry
Empetrum nigrum

Crowberries, often called mossberries or blackberries in some regions of Alaska, thrive in bogs, heaths, tundra and high mountains. They often grow tangled among the blueberries, and many berry pickers confuse the two. Crowberries are the blacker, shinier berry, sometimes growing in dense patches close to the ground. One of the most popular berries in the Arctic, crowberries often are overlooked elsewhere.

Harvest the mature berries in late summer to early fall. Crowberries are best after the first frost as the cold sweetens the fruit. If gathering them beforehand, place them in a freezer, which has the same effect as frosty nights.

WAYNE LYNCH

OUTSIDE PASSAGE

In 1940, after being separated from their mother for two years, 11-year-old Julia Scully and her 13-year-old sister Lillian boarded the spring's first boat to Nome to be reunited. These are her memories of that journey.

BY JULIA SCULLY

Julia [RIGHT] and Lillian Scully, before their trip up the Outside Passage from Seattle to Nome turned bad.

COURTESY JULIA SCULLY

EDITOR'S NOTE: In these days of regular jet air service, it is easy to forget that half a century ago Nome was most often reached by boat, and only after a hazardous 10-day voyage from Seattle. Julia and Lillian Scully's journey marked both an ending and a beginning. After Julia's father died in San Francisco, her mother, an immigrant with a fourth-grade education, returned to Nome, where the couple had begun their married life. She was forced to leave her children in orphanages for more than two years while she worked to establish a life for her family in Nome and at a roadhouse in nearby Taylor Creek. Scully's memories of the long journey that reunited her family are worth reading for the sheer beauty of her prose, and provide a vivid glimpse of how very different "coming into the country" was not very long ago.

IT WAS IN MAY—IT HAD TO HAVE BEEN May or perhaps early June, although I have no memory of time. At any rate, it was in May or early June that Aunt Sadie came to take us out of the Seattle Children's Home. I don't remember the leave-taking, if there was any, nor do I remember how we traveled from Queen Anne Hill to the dock where an Alaska Steamship Line boat lay at anchor. But I recall that she put us on board, gave us a box of chocolates, and then stood waving and dabbing a handkerchief to her eyes as the ship pulled away.

Lillian and I are not surprised to find ourselves on a ship. Neither are we happy or sad.

After the first day, we remain in our cabin, seasick.

The cabin is small and hot and smells of oil. The walls, white metal walls, vibrate with the thump, thumping of engines close by. I lie in the upper bunk and whatever I set my eyes on shifts, slipping away from my gaze until I close my eyes to shut out the dizziness, but then the room slides across the black screen of my eyelids.

Lillian lies in the bunk below me, breathing noisily through her mouth. Once in a while she says, "What time is it, Sissy?" As if everything would be all right if she knew what time it was. Or, maybe, she just says it to hear my voice, to see if I am still here, just to make sure. But, of course, I don't know what time it is.

I sleep and waken, not knowing how much time has passed, not knowing if it's night or morning. Only knowing that days have gone by in this room.

Sometimes a steward comes, stepping over the high threshold, balancing a tray, smiling, "A little consommé," he says. "Something easy, slides down easy," he says.

Smelling the staleness in that hot, close cabin, leaving the napkin-covered tray.

On the metal ceiling above me tiny cracks like tentacles wiggle out from a large rusted circle.

I could turn, turn my head away and look at the cabin, but it's better if I don't, it's better if I lie absolutely still.

Lillian and I hardly speak, but I know she is there, her steady, open-mouthed breathing in the bunk below a comforting background rhythm.

Once, we pull ourselves, shaky and pale and disheveled, from our bunks and up a metal staircase to the deck. And there's a wet, fishy mist, fresh and cool on my face.

And grown-ups there, strangers who seem to know us. "Rose Silverman's girls," they say, when they see us.

A boy my own age speaks to me, repeating what he must have heard the